TALES FROM JIMMY'S FARM

Also by Jimmy Doherty

On the Farm

A Taste of the Country:
A Traditional Farmhouse Cookbook by a
Very Twenty-first-century Farmer

A Farmer's Life for Me:
How to Live Sustainably, Jimmy's Way

TALES FROM JIMMY'S FARM

Jimmy Doherty

First published in 2022 by HEADLINE HOME
an imprint of HEADLINE PUBLISHING GROUP

1

Illustrations © bluebright, Svetsol, Fadhila Prasetyo Wibowo and iconicland, all Shutterstock

Photo inset pages: 1 bottom, 2, 3, 6 top and 20 copyright © Chris Terry
All other images courtesy of the author

Every effort has been made to fulfil requirements with regard to reproducing
copyright material. The author and publisher will be glad to rectify any omissions
at the earliest opportunity.

Cataloguing in Publication Data is available from the British Library

Hardback ISBN 978 1 4722 9291 9
eISBN 978 1 4722 9293 3

Publishing Director: Lindsey Evans
Assistant Editor: Kathryn Allen
Copy Editor: Jill Cole
Proofreaders: Anne Sheasby and Emma Horton
Indexer: Caroline Wilding

Designed and typeset by EM&EN
Colour reproduction by Alta Image
Printed and bound in Great Britain by Clays Ltd, Elcograf S.p.A.

Headline's policy is to use papers that are natural, renewable and recyclable
products and made from wood grown in well-managed forests and other
controlled sources. The logging and manufacturing processes are expected
to conform to the environmental regulations of the country of origin.

HEADLINE PUBLISHING GROUP
An Hachette UK Company
Carmelite House
50 Victoria Embankment
London EC4Y 0DZ

www.headline.co.uk
www.hachette.co.uk

For David Doherty and Christine (Kiki) Furney

CONTENTS

Introduction

GROWING UP

There was a certain inevitability to me having my own wildlife park. As a kid, I turned my dad's garage into a reptile house. It didn't matter – he could still use the carport for his van. Until I turned it into an aviary. Poor bloke, at least he could retire to his greenhouse. Oh, hang on, that was full of terrapins and Chinese painted quail.

When my parents moved from east London to the Essex countryside, I'm not sure they realised the effect it would have on their youngest child. The switch made sense. Dad, a builder in the East End, had spotted an old cricket pavilion complete with two acres of land, outbuildings and apple trees. The place had seen better days and was practically being given away. Mum wasn't quite so keen. This was well before the M11 was a glint in a road-planner's eye. Dad might have had designs on a little village outside Saffron Walden but as far as Mum was concerned he might as well have been proposing a move to the moon. Dad's argument was that London was too busy and crowded – you could get much more for your money in the countryside and have a far more tranquil life. It wasn't so much that he was a massive fan of rural living, it was more that he could carve a different kind of future out of nature. He could have his own plot, build his own house, raise his family – more *Little House on the Prairie* than *Emmerdale*. There's a picture of him cutting back an overgrown field behind the house with a scythe. It reminds me of when I first took on the farm and had to chop down acres of nettles and thistles, the difference being that at least I had a strimmer. Although, knowing Dad, I expect he put the scythe down the second the picture had been taken!

Scythe or no scythe, Dad knew with a bit of hard work he could wholly justify the upheaval of a move from London by creating the perfect family home. It's just that, albeit in a rather odd way involving West Ham United, the move triggered an intense obsession with nature in the infant me. My brother, Danny, who's six years older than me, was a big fan of the Hammers and so Dad cut him a little football pitch in the meadow grass. Clad in the latest kit, Danny would fire shots at me in goal. I also had some West Ham kit, but mine appeared to date from before the war. I soon got bored with this arrangement and while Danny dreamed of scoring the winner in the FA Cup Final, my own mind would drift away to wondering what was in the longer grass. I'd see a grasshopper and imagine what it would be like to be that small. If I was luckier, I'd see the odd scurrying mouse, even the occasional lizard. Without that distraction, who knows, maybe I'd have been the next Peter Shilton, but it was from that point that I started keeping and catching animals.

We had pets. Cats would appear and then stick around, each given the very imaginative name 'Puss', and then there was Duke the Dalmatian, who, like many of his breed, was affected by deafness. We'd call him but he'd never come back. Some hours later we'd receive a call from a constabulary in an entirely different county – 'We appear to have your dog – would you like to come and get it?' We also had a cocker spaniel, Sandy, with lovely long blond hair. I had a habit of giving him a comb-over or a centre parting. Looking back, he really was very good about it. But, even with the hairdressing, pets never had the same fascination as wildlife, although at the start they were a useful link between the two.

Initially, the cats did my catching for me as I gratefully collected the dead mice and voles they so generously brought into the house. I found them fascinating and assumed other people would too, taking them round to the retired wing commander next door. A lovely man, he and his wife tolerated me repeatedly bringing bags of dead rodents into their kitchen.

It wasn't that animals were cute – I wasn't cooing over things with soft fur and big eyes (maggots featured highly in my early collections) – it was total and utter fascination. I wanted to know how nature works. I looked at it like a mechanic – the bodywork is all very well, but what's under the bonnet? The result was I gathered up anything and everything. I was one of the few kids who, when asked if they were coming out to play, might reply, 'Not today. I'm hanging out with my owl pellets.'

The woods opposite the house offered yet more opportunity for investigation and adventure. I'd find myself alongside a little river with a scrabbly bank and lots of twists and turns. It was the tiniest shallow watercourse, but in my head it was the Amazon. I would paddle across it, thinking of the creatures I might be encountering in the watery depths of South America. Forget birthdays and Christmases, Mum and Dad's decision to relocate was the greatest gift they could have ever given me. Whereas in another life I might have been kicking a ball in a back alley, here I was absolutely captivated by the wildlife around me. I had tunnel vision. Over and over I was happy to lose myself down that hole.

Occasionally, there'd be school trips to Colchester Zoo, where I'd wander from zone to zone suitably impressed. What I really wanted, though, was my own live animals. I began with standard fare like mice and stick insects, and, early on, my mum gave me a budgie that a friend's daughter couldn't look after. But I wanted to expand, which was how the pond, soon alive with frogs, toads and fish, came about. Dad also built me a small vivarium – a brick reptile container with a glass front – which meant, excitingly, I could expand into the non-native side of the animal kingdom.

It was now, bit by bit, that I began to take over areas of the property. Stealth was my method of choice. I made the generous offer to Dad to clean out the garage. With it empty, I set about painting the entire interior white. It took me forever. When night came, I'd paint by torchlight with the door shut – it was a wonder I wasn't overcome by the fumes. I threw in a bit of old carpet and, with the makeover complete, filled it with vivariums in which soon resided more exotic fare – like snakes. These doubled as a very good way of keeping my hard-earned pocket money safe. I put a jam jar full of notes and coins in with my North American water snake – similar to a grass snake but which can be quite aggressive if it feels threatened or frightened. Of course, you have to be pretty comfortable around snakes to do this, otherwise you'll never get the jar out again.

Polecats, essentially wild ferrets, as opposed to those we domesticated hundreds of years ago, were another firm favourite. In the wild the polecat has a black mask, while the domestic ferret has fewer markings. Mine were a cross between the two. My mum and nan had long used a phrase about 'stinking like a polecat'. I'd never understood what they were on about . . . until I kept them. Polecats, like skunks, are mustelids, kitted out with anal scent glands for marking territory and signalling to the opposite sex. They really do smell bad, somewhere between rotting food and a public convenience. A word of advice: don't keep one if you're trying to sell your house.

Odour aside, I was fascinated by them. I could interact and play with them, and yet they maintained a real wild instinct. They really are incredible hunters and if one bites you then you really do know about it. That tended to happen at feeding time when they'd grab day-old chicks from my hand, sometimes a little too keenly. Those chicks were bought frozen from a supplier. They then needed to stay frozen, a fact which would quite often prompt a small shriek from Mum when she reached for the fish fingers, only for her hand to alight on an entirely different inhabitant. If I was freshly restocked there could be several dozen in there.

To be fair, I had a keen interest in live birds too. I'd fancied an aviary ever since I'd accompanied Dad on a building job when I was eight. His client had a fantastic birdhouse and I was obsessed with it, same as I was with the one I used to see in the Harvey Centre shopping mall in Harlow. Both those aviaries were professionally constructed. Mine was put together with bits of wire from skips and odds and sods I'd found among Dad's building gear. Wildlife meets *Steptoe and Son*. I tied branches to the inside, added some plants, and headed off down to a stockist. I walked in to find a vast room, a wonderland, full of exotic birds. Somehow I managed to make a choice, leaving with some finches and cockatiels, and releasing them into what was actually a good-sized aviary. I added my growing collection of budgies which, until then, could only take wing in my bedroom. To see them and the other birds fly around in the aviary was amazing. I'd lie on the floor and watch them swoop over me, loving the fact that I could hear the garden birds add their own voices to the chorus.

My enthusiasm at building enclosures far outweighing my ability, I came home from school one day to find all my budgies sitting outside in a tree. I couldn't help noticing they looked exceptionally pleased with themselves. My mates watched laughing as I shinned up the bark to retrieve them, stuffing them unceremoniously into my pockets. I got back to the aviary with the budgies just in time to see my beloved Senegal parrot flying off into the woods. My pals found this exceptionally

8

amusing too. I'd wait until they were least expecting it to carry out my retaliation, dropping a snake in their lap and watching their terror unfold.

Being a normal sort of child, my next job was to line the greenhouse with bubble wrap. The idea was to keep the warmth in, but essentially the effect was to create something half-greenhouse, half-spaceship. I banked up a load of soil against the inside walls, and then, using some of Dad's cement, made a terrapin pool in the middle of the floor. They lived happily alongside a few Chinese painted quail, which also favoured the toasty conditions. Chuck in some tropical plants, butterflies and a few stick insects and there it was, my own tropical house. OK, I could barely move in there but it certainly did the job. And my dad did eventually find somewhere else to grow his tomatoes.

If that's how I treated the areas of the house that weren't my own, it's hardly surprising that my bedroom was, fair to say, crackers – posters, books and animals everywhere. Early on, I'd noticed that any wildlife collector worth their salt had an aquarium – I just needed the £27 to buy the one I wanted. Luckily, I had a little ruse going with a mate. We'd go round the village offering to clean cars – £1.50 for a wash; £2.50 for a wash and wax. Bargain. Except we'd only do the side facing the customer's house. Good news was I'd cobbled together precisely £27 before the game was up. Not long after, I returned to the house with a large glass tank. I hauled it into my bedroom and set about filling it with all manner of fish. When, finally, my dad saw it, all hell broke loose. 'Get it out the house!' he ordered. 'It'll cause damp.' He calmed down after a while, possibly consumed by the question of how on earth I'd managed to get it up to my room without him noticing. My Uncle Ron was a bit more understanding. He was a keen angler and kept tropical fish.

In all honesty, the fish should have been the least of Dad's worries. I also shared that room with spiders, toads and snakes. On one infamous occasion I went abroad with a friend for a few days, with my nan staying over in my bedroom while I was away. Naturally, that just happened to be when my royal python made its bid for freedom. It spent the next few days settled in the nearest 'branch' it could find – my curtain rail. Fortunately, Nan never clocked its presence, even though she would have been standing at my window with the snake just inches from her head. I should point out that these beautifully patterned creatures are actually non-venomous. Coincidentally, we now have a royal python at the wildlife park. Fear not, it's never yet been found on a curtain rail.

Thing was, I saw that space not so much as a bedroom but as a jungle. I'd turn the light down really low, put a tape of jungle sounds on, crouch among the potted plants and pretend it was a rainforest. Sounds mad, but I'm sure we've all lost ourselves in our imaginations at one time or another, be it while playing with soldiers on the carpet or riding an exercise bike while watching the Tour de France. We are all capable of escapism – it's just that some go on to make it more than that. To do so, though, you really do need an intense interest in and passion for your chosen subject. True to say that if I didn't have that complete all-consuming fascination with nature, no way would I be looking out at this wildlife park now.

It wasn't just doing, it was reading. I still have a lot of the wildlife books I read as a child, including Gerald Durrell's *A Practical Guide for the Amateur Naturalist*, the bible for anyone wanting to watch and understand the natural world. It gave tips for discovering all kinds of animals and habitats, on the back of which I built a collection of kit – penknife, magnifying glass, butterfly net, etc. – that I'd take with me on expeditions. In my head I'd be Indiana Jones. Once we visited an uncle who lived in Spain and I happened across a shop that sold the perfect Indiana Jones accoutrements, a leather hat and a whip – not toys, the real business. When I got back home, I couldn't wait to don my new gear, even adding a little leather satchel, and heading out on adventures. I'd see a wood a mile away in the distance and that would be my target. Sounds a bit random, but when you're nine that kind of thing is brilliant. You get there, find somewhere to sit down, have a jam sandwich, and off you go. I had visions of entering ancient cave systems, potentially being chased by a huge round rock. More likely I'd have a go at cracking the whip, get it too near my face, split my lip, and have to go home.

Adventure films such as *Raiders of the Lost Ark* opened a window into a different world. They encouraged me to go out and investigate what was around me. Even now, wherever I am, I love that idea of just heading off in a certain direction and seeing what pops up along the way. I suppose it could simply be seen as an over-active imagination, but I don't mind because it makes life more interesting. That's why the Natural History Museum always appealed to me. I never saw it as just glass display cases, dry collections of artefacts. What I saw was the story of how each particular item got there. If it was a tropical butterfly, I'd be thinking of an explorer going up the Congo in a little wooden boat. If it was a fossil, I'd be imagining someone digging up a lump of rock and seeing the outline of that ancient dinosaur. Sealife was the same. All I could think when I saw an amazing and intricate shell was: 'Imagine seeing that on the bottom of the ocean for the first time. That must have been incredible!'

I had an atlas of wildlife, each chapter a wonderful illustration of a country or continent's wildlife, and I'd get lost in the thought of going to those places and witnessing with my own eyes the animals I'd seen in the photos and illustrations. I daydreamed about heading to North America and seeing a grizzly – preferably from a distance! The more I saw, the more I felt compelled to travel and view animals in their native environments, which, in later life, I was so lucky to be able to do.

I was thirsty for knowledge. My shelves were full of books about fish-keeping, ferret-handling and butterfly-collecting. Then there was the A–Z of tips for owning a snake. I was fortunate also that there were some real landmark TV shows. Gerald Durrell's *Amateur Naturalist* wasn't just a book – I'd get off the school bus and run home to watch the veteran conservationist catching a lizard to show off its remarkable patterning and features. Another series which had me hooked was *Supersense*, all about how animals perceive the world, their heightened senses and how they use them to survive. The way the show portrayed the inbuilt intelligence of animals without constantly comparing them to us, a whole different way of looking at the subject, really resonated with me.

At its peak, my domestic menagerie featured tortoises, terrapins, snakes, lizards, cockatoos, guinea pigs, barn owls, rabbits, parrots, fish and ferrets – to name just a few. I became quite well known. It wasn't unusual for the village cub pack to turn up at the door wanting to take a look, and the parish magazine did an article on me. 'That's it then,' I thought, 'I've made it!'

Considering the level of my infatuation and its effect on their surroundings, my parents were very understanding. They'd forever be looking for a pot that had mysteriously vanished from the kitchen, while I'd be sitting there tight-lipped knowing it had been requisitioned as a home for a praying mantis. My dad would go to use his wheelbarrow only to find it filled with water and half a dozen terrapins. He'd open the door to his shed and find it full of rabbits. I think they saw it as a phase I'd grow out of. Hmm . . .

What Mum and Dad did like was having a child who had an interest, as opposed to one just kicking his heels. I wonder if, from them, I subliminally inherited an ethos that doing, rather than sitting back, was what got you somewhere. They did, after all, both run their own businesses, and never shirked hard work. I spent a lot of time organising, setting up, drawing plans – I had a real love of detail. I think actually they enjoyed helping me out. But in the long term they were concerned as to how I was going to make a living out of a love of nature – that basic question of what job it could lead to. Totally understandable. They were part of a generation where you got yourself a trade or went to work for someone else.

Dad was a funny mix. He would see the negative in me having an animal – 'If you keep chickens, they'll attract rats' – while at the same time building me a tank in which to keep my latest purchase. He'd glue panes of glass together and add a wooden lid – perfect for my garter snakes; or construct a hamster cage like a miniature rabbit hutch. He'd see me making something and try to help me out. Where it might all lead, though, he couldn't imagine. My brother also watched on bemused. He couldn't have been less interested in wildlife. His only link to me and animals was leaving the cat's bowl near me when I was a toddler in the hope I'd grab a handful and eat it – which, apparently, to his delight, I more than happily did. The irony is he now works at the wildlife park, where he puts his building skills to good use. All the stuff I used to try to construct, he can actually do properly. As I mentioned, he's six years older than me, and I'm sure he hasn't forgotten quite how irritating I could be as a child. He'd have been well within his rights to have strangled me from time to time. He liked Lacoste clothes – an unfortunate choice, as Lacoste has a crocodile logo. One of my most annoying habits was to tear the logos off his shirts and stick them on mine.

•

The wildlife infatuation didn't begin and end at home. At school, me and a couple of friends produced our own wildlife magazine – *The Natural Choice*. We made it look as professional as possible and filled it with articles about the rainforest, wildlife of all sorts – anything related, as the title would suggest, to nature. We'd reel off a few dozen copies on an old turn-handle printer and hope we had a bit of an audience. I've a feeling the distribution figures reached no higher than a few of the teachers, but again it emphasises how even early on I was really keen on letting people know the importance of wildlife.

True also to say that without my interest in nature I wouldn't have cared in the least about school. Unsurprisingly, considering I spent so long in another world – be it minute after minute staring at my fish or hour upon hour peering into my snake tank – academically I wasn't brilliant. Nature made sense, everything else seemed so abstract. Trigonometry – who cares? But give me information about aviary birds, or keeping tropical fish, or snakes, then my hunger was insatiable.

That disaffection with school was further emphasised when, at the age of just thirteen, I began working at an incredible place, a nearby farm and wildlife park which went by the rather wonderful name of Mole Hall. It wasn't quite as easy as just walking in there and starting to groom the nearest monkey. I did have to attend an interview – I was mortified when my mum decided she'd attend too. I hoped she might stay in the background but actually she was extremely vocal and answered most of the questions. Mind you, she must have given the right answers as I ended up getting the job!

Subconsciously, I suspect, Mole Hall has been a definite influence on what we've achieved here – we share a lot of characteristics in terms of ethos and animals – and for that I can thank Pamela Johnstone, Mole Hall's owner and a truly amazing woman, sadly no longer with us. I'd hang on Pamela's every word as she chatted to me about her trips to Africa or South America, and listen to her ideas about combining farming and conservation with a wildlife park as well. There was a massive age gap of half a century between us, but it didn't matter in the slightest because we were both fascinated by a timeless subject – the natural world. The idea I saw brought to life there – if you can make a difference, you should make a difference – has never left me. In fact, once Jimmy's Farm started to expand into new areas, a friend actually said to me, 'You've recreated Mole Hall Wildlife Park.' I had to reflect he was right. In many ways I have!

One way or another, be it part-time, full-time, in school holidays, or between terms at university, I was at Mole Hall for ten years. It made such an impression on me, like landing in heaven for a lad like myself, and so thrilled was I to be involved that I can still remember how much I earned at the start – £1.13 an hour. Not bad, you could get a good few Curly Wurlys for that back in 1990. In the summertime, I would help on the neighbouring farm with all sorts of bits and bobs, but mainly – and this is where the excitement was – I was based in the tropical butterfly house. I can still vividly recall that first experience of opening the sliding door and the heat and humidity hitting me in the face, seeing tropical plants and the most stunning butterflies everywhere. It was like finding heaven. Now, of course, I have my own butterfly house, but every time I open the door I'm still taken back to being that thirteen-year-old kid, so awestruck, dumbstruck even, by that first time at Mole Hall. I look at our young visitors here and wonder if they're having the same experience. Is it something that will stick with them, that might make a difference in how they see the world, or the direction they want to take in their own lives?

You will notice throughout this book that I'm very susceptible to being transported back to childhood memories. It doesn't have to be something incredible, the slightest, most banal thing can send me back to days in short trousers – like having a hot bath. As soon as my toe hits that water my brain thinks just one thing, 'Oh no! I've got a spelling test in the morning!' (Sunday was always bath night when I was a kid.) I think it's because the young me and the adult me are so intrinsically bound together. In many ways I still have the same brain, and I'm pleased about that, because if I didn't, I really don't think I'd be doing this now.

People often ask me where I sourced my childhood collection. These were, after all, the days pre-internet. Strange now to think that a good deal of it came from the pages of *Exchange & Mart*, an odd publication ostensibly full of cars, motorbikes and bits of machinery. Flick to the back, though, and you'd find people breeding and selling all kinds of animals. Then there were the specialist publications, such as *Practical Fishkeeping* or *Cage & Aviary Birds*, again with endless classified ads as people offered various species for sale. With these magazines under my arm, I'd gather a pile of ten-pence pieces and head down to the village phone box to put in some orders. Because these were live animals it had to be next-day delivery, which cost a fortune, so to make it viable I'd put in large orders just a few times a year. Next day, as promised, a delivery van would turn up. The driver might have carried the box to the door a bit quicker if he'd realised there was a snake lurking within. That whole pet trade has changed drastically in the years since, and for the better.

Alternatively, an animal would occasionally be gifted. My cousin Ricky once bought me a bearded dragon, handing it over at my mum's hairdressing shop in Stratford, where, as a kid, I had to go on Saturdays, sitting there while a stream of old ladies had their hair permed. I kept the 'beardie' in the back room and then took it home. It's just how I was. Relatives would often ask what I wanted for my birthday or Christmas and be met with the answer, 'Money, please – I'm saving up for a python.' Chances are when Santa asked what I wanted, I replied, 'Your reindeer.'

It occurred to me after a while that if I had an excess of a certain animal I might myself be able to make some money by moving it on. That way I'd be able to indulge my love of wildlife for much longer than just school. I used to take animals into the classroom occasionally and for a while had a solid customer base who would buy a rabbit or a snake or whatever. I was like a *Beano* character – Jimmy, the kid who always has something odd in his pocket. I'd put my hand in and forget there was a toad in there. I'd have made a great wizard. Although not many wizards have ever travelled home from sixth form with a snake in a sack.

On one occasion, I had some black mollies which bred and resulted in a bit of a glut. I was just thinking how I could sell them, albeit for just a few pence each, when, to my horror, my Siamese fighting fish woke up. It took one look at these tiny black dots and started eating one after the other. As I desperately fished as many out as possible, it occurred to me that I'd made the classic mistake of trying to catch the prey rather than the predator.

I'd breed birds all the time. I had three chickens when I started, and ended up with fifty-three. Trouble was, they'd disappear into the undergrowth and I couldn't find their eggs. Soon after, they'd reappear with a trail of ten chicks scurrying along behind them. My budgies were also forever reproducing, as were my zebra finches. More than a chance of being able to fund a career in wildlife, what I took from the situation was an appreciation of, and extremely rapid education in, the life cycle – the next stage up from watching tadpoles develop in a jar. I learned so much more through first-hand experience than from reading a book. But what reading books and studying biology at school did teach me was the technicalities of what I was witnessing – the whys and wherefores. Combine the two and I was receiving an incredible injection of know-how.

That close-up education also removed any fear. I was drying my daughter's hair the other day and an earwig ran past. She's scared of them, same as my wife Michaela's scared of spiders. At times like that I always say, 'Look, it's not out to get you. It just happens to be sharing your space.' Often we remove ourselves so far from the natural world that it becomes a threat. I was once on a beach in Fiji with a storm approaching, and around forty yellow-lipped sea kraits, a venomous sea snake with a neat trick of fooling predators by having its tail look like a second head, emerged from the waves and onto the sand. People were absolutely terrified, screaming and lifting their children into their arms. But this was absolutely normal behaviour for the species. The snakes were simply heading to the beach to shelter and rest. They couldn't have been less interested in the humans who happened to be there – the yellow-lipped sea krait eats small fish, not people. I think it's very important for kids especially to understand that wildlife doesn't exist as a threat. It's not relating to us in that way at all. We don't need to be stand-offish. It's for that reason that I let my kids, once they understand how to handle and not damage them, catch butterflies in the garden. That close-up appreciation is so important. They see the delicacy, the colouration, the biological detail. Do that just once and you'll never look at a butterfly, or any other creature, in the same way again.

Every year, I looked forward to going abroad on holiday. I'd see a gecko running over rocks and up walls and think it was magical. I'd chase them, my jelly sandals no match for their nimble grip. I went to Greece with my own children recently and watched as they were mesmerised by exactly the same thing – the lizard's colour, how it uses its feet to climb. We caught a praying mantis at one point and observed it for two days before releasing it. I still absolutely share a childlike excitement at that sort of thing. I remember a holiday in a tiny villa with my mum and her sister where I spent hour after hour staring at these ginormous ants – well, to me they were ginormous! The ants, the heat, the palm trees, the noise of the cicadas, was utterly beguiling. It didn't matter where we went, home or abroad, I'd always find something to explore. I'd spot a reptile on a scruffy patch of land, or lose myself messing about with a dead jellyfish. I can still remember the amazement of putting goggles and a snorkel on and being able to watch fish in their natural habitat rather than in a tank. I was always going to learn how to scuba dive after that.

One year we went to Canada, a trip which had a huge effect on me. We visited what was then the Metropolitan Toronto Zoo, an absolute pioneer, grouping animals by their zoogeographic origins and placing them in large, naturalised enclosures rather than cages. The detailing extended to expanding the correct terrain and vegetation to outside as well as inside the enclosure. The effect was deeply immersive, and I revelled in it, even getting, albeit a little tongue-tied, to speak to some of the keepers. I told them about my job at Mole Hall but somehow it just didn't seem to compare. Toronto might be 3,600 miles away from here but there is definitely a little bit of the old Metropolitan Zoo ethos in what we've tried to do. I love the idea that you can build a wildlife park where enclosures fashioned from existing environments slot so seamlessly together that they just become part of an ever-changing landscape. Visitors should never find themselves walking from one concrete bowl to another, staring at its inhabitants and moving on; that old attitude of 'big cats over there, elephants down here, penguins in the middle' that seemed so often to happen.

Look at photos of how zoos used to be and you realise it's not actually that long since they were putting chimpanzees in clothes and wheeling them out for a daily tea party. Alternatively, visitors would be offered the chance to feed the bears or ride a camel. It almost felt as if zoos and circuses were one and the same. Thankfully, the days of humans being entertained at the expense of the animals is, at least in this country, now gone.

I knew right from the start that I never wanted our wildlife park to be a place that has elephants, lions, giraffes, polar bears, tigers, those traditional iconic 'zoo' greatest hits. I wanted animals that were here for a reason, and to provide them with as big an area as possible so people could watch them in the most naturalistic way. A modern wildlife park's environment should reflect its purpose – to educate about the importance of biodiversity while performing vital roles in conservation and protection.

It's funny how, as with the Toronto trip, the past and the present can link. As a kid, when I wasn't reading I was drawing. I'd imagine different worlds only to be drawn back to one fantasy in particular. Hour after hour I'd draw my ideal wildlife park. I don't mean rough sketches. Enclosures, animal houses, water features would all be planned in intricate detail across A4 sheets of paper taped together. I could get very geeky about that kind of stuff. Actually, thinking about it, I haven't changed!

Writing this book prompted me to take a journey back to those pictures in my head. I wondered if what I imagined then bore any resemblance to what I see in front of me now. Give or take the odd raccoon, there is, I'm sure, a resemblance, albeit one which came about more by accident than design. Twenty years ago, when I first set eyes on this site – derelict, overgrown, dilapidated – it was as if I'd been given a big block of stone. I had an idea what form I wanted to sculpt it into but was fully aware that my lack of know-how meant that along the way there was a good chance I might chip off a nose and have to turn it into an ear. With any venture of this kind, you're ever evolving. You have a plan and add. Nevertheless, I like to think that if my childhood self turned up tomorrow he'd be at least mildly happy with what he saw. I just won't be letting him build any enclosures. I'd quite like our animals to stay out of the trees!

I may have moved on a little, but I still look at the wildlife park through a child's eyes, particularly how we exhibit the animals. I think about how they will behave within their enclosure and how that will affect the experience of the children. The new meerkat enclosure has a big window. It might sound obvious but if I'm a kid, I want that window to be low enough for me to be able to see in easily. I knew as well that it would be a good idea to have some logs by the glass. That way the meerkats would be attracted to that part of the enclosure to sit on them. As a kid, I want to be excited. I want to see the meerkats from a distance and then get right up next to them, to feel as if, were it not for the glass, I'd be able to touch them.

It was the same when we talked about a new butterfly house. I thought it would be a good idea to make the entrance dark. That way, kids would emerge from jet black into an incredible burst of colour as these amazing creatures danced around their heads. I like to create that element of adventure. 'Edutainment' they call it – learning not from a rigid biology lesson but through discovery. It's a way of informing and educating by stealth. If someone, child or adult, takes a look at Basil, our anteater, they'll see he's got his bedroom and his private garden. But then they'll see the tunnel that leads out to the paddock, and then the other tunnel that leads to his own bit of woodland. That's great for Basil – and what kid doesn't like a tunnel? – but it's also a way of showing that Basil is happy in different habitats.

My old habitat, that house, that old cricket pavilion where I lost myself in wildlife, is now itself lost – demolished. But it will always live on in me. I still feel an intense connection when I see the fields where I grew up, the spot where Dad planted his apple trees, the old pond where I used to keep terrapins.

Right there was my little jungle – but I'd end up with something rather bigger.

My brief intellectual look – inspecting butterflies at the Natural History Museum.

Very few things a pig likes more than a tickle behind the ear.

The only thing that was static in those early months was the caravan.

For ground clearance, look no further than pigs or goats.

If it wasn't for the ear tag, this could be the 1930s, not the 2000s.

As the years went by, we increased engine capacity.

On your way pal! Geese can be very territorial.

I'm so lucky to be part of a landscape that could be mistaken for a painting.

A pink tractor on our wedding day – whoever said romance is dead?

'I won the droopy tights competition!' Molly and Cora
– Michaela wasn't taking part.

STARTING OUT

Endlessly winding lanes – that's all I seemed to see for a few months in 2002. I'd had an idea that I'd like to be a pig farmer – doesn't everyone? – but finding the right place was an impossibility. And then one day, with no real hope or expectation, I came to Pannington Hall Farm. Parking up, I walked up a grassy track and into the heart of the farm. The first thing I saw was an old derelict barn, tin roof, trees growing through the walls, their roots ripping up the floor and destroying the tiles. Amidst all that devastation – natural devastation, plants and animals inhabiting the brickwork, reclaiming a space that humans had left behind – there was a single wonderful rose poking its head out of a thicket. It somehow represented the whole place. Where most people saw a bombsite, I saw perfection. All I could think of was the opening sequence of *The Darling Buds of May* – that idyllic sun-swept golden country image. Romance and atmosphere, which I felt was vital if I was really to invest myself in a place, was here in absolute abundance. I didn't want somewhere that felt modern, routine, workaday. If I was going to head down this route, I needed a place that meant something more to me, that would lift me and inspire me to achieve what I wanted. That was an absolute prerequisite, because making this happen was going to take an awful lot of hard work. From a distance everything appeared quite rustic, and I'm sure it would have made a lovely watercolour, except wash away the paint and the canvas underneath was horrific. In practical terms, there was no water, no electricity, and every building I looked at, including the Grade II-listed sixteenth-century farmhouse, was knackered. Pannington Hall Farm had been a dairy farm, but the upsizing of agriculture had made running a small herd less and less viable, and in the end it had become too tough a business. The farm had fallen into rack and ruin. The entire place had been falling down for twenty-five years. With a task of that magnitude, it's passion and imagination that drives you through.

Some people might see dreams and reality as being concepts alien to one another in the business world but, in all honesty, I doubt the wildlife park would have ever happened if I didn't have that element of romanticism that so fired me on that first visit. Look at a project like this on a spreadsheet and it's about as romantic as a long weekend mucking out a cow byre. But I don't do it for the spreadsheet. At the end of the day, when my time comes to shuffle off this mortal coil, it's the romance I'll take with me, the incredible memories of family, friends, and a thousand and one animals, not row upon row of dry facts and figures.

The other thing I had on my side was naivety. There's a saying about a little bit of knowledge, but actually I think it's powerful rather than dangerous as it allows you to venture into areas from which you might otherwise shy away. With full knowledge, you'd never make the move. After all, I set out with a vision of giving acorns to pigs on a spade, not, as it soon turned out, working eighteen-hour days covered in nettle stings and horsefly bites. Naivety papers over reality and allows a dream to be pursued. A more grounded and pragmatic person would have seen Pannington Hall Farm and run a million miles. Straight away they'd have spotted the hurdles hidden in the undergrowth, the issues that would soak up cash like a sponge. They'd have seen a venture dead on its backside after a few months. But in doing so they'd have blinded themselves to opportunity. Truth is, opening yourself up to new, occasionally mad, ideas is a way of creating sustainability. It might not be on the same scale – not everyone has a resident anteater – but what we've done here epitomises the revolution that other rural businesses have gone through. More than ever, they're saying, 'Don't drive past – come and see what we've got – what we do.' Encouraging visitors to see, taste, smell, take part is important now. And it's working.

Naivety in all its forms should be embraced. It's the very challenges it throws up that ultimately provide the greatest memories. The days when it all felt like it was going wrong but somehow you pulled it out the bag. Or the days when it went wrong and actually stayed wrong but somehow you staggered back to your feet and carried on. It allows you to learn from your mistakes. I remember a programme, *A Farmer's Life for Me*, I once hosted on BBC2. The idea was that couples, hoping to escape the nine-to-five, competed to run a 25-acre Suffolk smallholding rent-free for a year. As they were set tasks ranging from identifying a bounteous plot to selling a strawberry harvest, the ones who made it furthest through the process had the humility to admit they didn't have a clue about a lot that was going on. They were also the ones most willing to learn. Sometimes an empty sheet of paper isn't a bad thing.

Time, place, ambition, naivety, desire – each and every star was in alignment. Time especially – had I taken the leap any sooner I'm pretty sure I'd have been too juvenile. I was fairly juvenile as it was! Too many years later and I'd have been too knackered. It was the Goldilocks moment. Even so, it took a few people aback when I signed on the dotted line, Mum and Dad included. They thought the idea was madness and wouldn't last, and not without good reason. I had a track record of diving head first into something one minute and being bored with it the next. The only area where that wasn't the case was wildlife, but it had certainly been true with learning a musical instrument and one or two other ill-fated hobbies. But where others saw only a lifeless wreck sunk by the cold intensity of modern farming techniques, I saw buildings I was convinced must once have bustled (the dilapidated barn, I later discovered, was once the longest thatched building in Suffolk) and a hundred rolling acres that would have been abundant with farm animals. I was actually standing on ground recorded in the *Domesday Book* of 1086, a point when it was in the ownership of Swein (not swine!) FitzRobert of Essex, son of Robert FitzWimarc, a relation of William of Normandy and Edward the Confessor, who, strangely enough, was based predominantly in my old home village of Clavering – maybe all this was meant to be! It was written in the stars!

The land is also believed to have once been owned by the Augustinian Priory of St Peter and Paul, and later, in the 1500s, by Sir William Butts, the King's physician, serving in the court of Henry VIII. The Butts clan were prominent Puritans, so I'm not sure if they'd have enjoyed our sausage and beer festivals! In more recent history, there was a Second World War anti-aircraft battery here, while the Women's Land Army, better known as the Land Girls, made sure the cows were milked. I didn't know the history of these fields at that point but I've always been glad since that in our own way we are adding to them.

Pannington Hall Farm might have seen better days, been on its knees even, but it wasn't ready for the chop just yet. In fact, the beauty of a tumbledown farm that had become home to a multitude of nettles and weeds was that its ramshackle nature meant a rent I could afford. All the 'oven-ready' spaces I had looked at were parts of existing farms, whereas this seemed ideal in every way, a great mix of woodland and pasture. The fact it was self-contained meant I could do pretty much what I wanted. Not only that, but its location just off the A14 meant London and Cambridge were within easy reach, as were a whole load of little towns we could sell to when, as I planned, I started marketing meat and sausages from our intake of rare-breed pigs.

*

I'd come to Pannington Hall Farm by a typically circuitous route, one that might have ended before it had even begun. Aged sixteen, I sat down to decide what I wanted to do. I wrote 'Royal Marines' on one side of a piece of paper, 'Zoology' on the other, and drew a line down the middle. I then filled each column with the potential benefits. There might appear to be a cavern between natural history and the military, but actually to me there was one huge and significant link – order and classification. They are quite obvious bedmates. It was neck and neck but the fact I was already working in a wildlife park and by now volunteering at the nearby Saffron Walden Museum, helping with their natural history collections, swung it for me. To still get that taste of military life I joined the Territorial Army (TA), spending five years in the Royal Signals. Sartorially, a career wearing civvies was definitely the right choice. Army fatigues don't suit me – green makes me look washed out!

Earning a few quid from the TA, as well as working at Mole Hall, stints on the checkout at Tesco, plus pot-collecting and washing-up in The Cricketers, the pub owned by my mate Jamie's dad, meant that as well as volunteering at the museum, I could hit the shops in Saffron Walden too. Jamie would be buying drums – at that point he thought his future lay in being a rock star – while I'd be handing over my hard-earned in exchange for lizards or snakes. In a vague attempt to look fashionable on these trips, I'd put so much mousse on my hair that it quite literally wouldn't move. I'd be there in blue stone-washed jeans, blue striped shirt and blue boating shoes. Jamie, meanwhile, had green stonewashed jeans, a green striped shirt and green boating shoes. I had dark mousy hair, he had blond mousy hair. I wore Kouros, he wore Blue Stratos, applied in such quantities that we both stank – a nod to Grandma here – like polecats. The girls could smell us from miles away, the Pepé Le Pews of our day, although we can't have been too bad – he first met his future wife, Jools, when we were teenagers on a double date. By that time I'd already known Jamie for well over a decade. I still fondly recall standing on the back of his cloak and strangling him after we were ill-advisedly cast as two-thirds of the three wise men in the school nativity play. Now here we'd be, sneaking into the pub next to the police station for a couple of drinks. To be fair, we weren't the only ones. I think the police were OK with it – at least they had all the under-agers contained in one place!

In some ways then, I was a typical teenager. But in so many others I absolutely wasn't. For instance, while most teenagers spent their money on clothes or the latest bit of electronic gadgetry, my wallet tended to open for fish tanks, a field in which there was fierce competition. A simple goldfish tank was like having a basic Mark I Ford Escort, while a tropical tank was a step up, the equivalent of an Escort RS Cosworth. A tropical marine tank? Now we're talking! That was the fish tank equivalent of a Lamborghini.

The contents of a tank also mattered. I was forever heading over to a tropical fish shop in Hatfield. At one point I had a lionfish and a stone fish, both of which are venomous, a real 'Wow!' moment among the local fish community. I'd always go for the most unusual fish and also built up a really good collection of tetras – neons, cardinals, black widows – with an incredible array of colours. It was then I spotted the shovelnose catfish. I'd never seen one of these deep-river fish from South America, and it was amazing – long, slender, with striking spots and stripes. Priced at £22 it was big money, the equivalent of a bottle of Kouros, the other altar at which I worshipped. But it was a no-brainer – I had to have it. I was so, so proud of my new purchase and couldn't wait to get it home. I slid it into the tank and, in a sorry repetition of the Siamese fighting-fish incident, it hoovered up every single fish I already had in there. It then looked at me with a huge, distended belly and swam off. From that point on, when anyone asked me about my collection, I used to say, 'Collection? Haven't you heard? I just keep shovelnose catfish now. I'm not bothering with anything else.'

After taking my A-levels, I took a year out. I travelled across the world, at which point all those multitudes of wildlife books I'd read and the pictures I'd seen became a reality. For the first time, I saw rainforest, but what really changed everything for me was diving in the South Pacific. I'd see coral reefs teeming with sharks, lagoons full of octopuses. I lugged an enormous backpack with me wherever I went, filled mainly with a camera and dozens of cassettes for my Walkman. I had very little room left for clothes. To me, though, that didn't matter. I wasn't there to party. As ever, it was much more about seeing landscapes, nature, a world without lots of people.

I did take a notebook with me, which I happened across recently. I think perhaps I was bordering on a hippy phase because it was full of reflections about us being mere grains of sand on the beach of life – the kind of stuff that spouts so easily from our mouths (well, mine anyway!) at that age. I work with college students occasionally and cringe when I recall some of those things I wrote. Just the other week I was talking to a group about how to write voiceover material for their YouTube channels. They were amazing young people but as I looked around the room I couldn't help thinking, 'I'm so glad I'm not a student any more!', that whole thing of finding your way, being so unsure of yourself. It reminded me that it's a period of life that can be really quite difficult. I also couldn't help feeling, as I explained about the workings of TV, how much I must sound like Alan Partridge – who no doubt they were way too young to have heard of.

Back home after my international trip, I completed a degree in animal biology at the University of East London, amazed to find that the campus, in Stratford, was directly opposite my mum's old hairdressing shop – she had changed career to reflexology. Going to a university within reach of home meant I could still keep my job at Mole Hall, so I had the best of both worlds. It was obvious, though, that in the coming years I'd be spending less and less time at the house and so I began paring down my own mini wildlife park. Some animals went to other collectors, others actually ended up at Mole Hall – I still vividly recall the moment I let my finches fly free in the butterfly house.

I stayed on at Mole Hall for another two years after I graduated. I lived in Cambridge at the time, and – old habits die hard – still kept a few animals. At one point, I had some Atlas moths hatching out in the spare room. Endemic to Asia, this particular moth can be bigger than a man's hand. When a friend came to stay, I directed him to his room for the night and thought no more of it. In the pitch-black depths of the early hours, he was woken by a flapping of wings. He then felt something land on his stomach and start walking up his chest. It stopped at his neck. Understandably perturbed, he turned on the light to find himself wearing a living bow tie. Like most visitors, he saw my digs as a house of horror. The kingsnake in the bathroom tended to confirm that view. I liked to watch it while I was having a bath, but other people didn't seem to find the presence of the constrictor quite so relaxing.

For a while, I worked in the Natural History Museum's entomology (the study of insects) department.

I then headed to Coventry University to study for a PhD in the specialism. I had a lab in the basement where, while building an intense fondness for pork scratchings (perhaps a sign of things to come), I bred insects, just along from Mark, the animal technician, who looked after the snakes and reptiles used for teaching the biology of cold-blooded animals. Occasionally, Mark and I would go down to the river by the M6 motorway and catch American signal crayfish. An invasive species, introduced in the seventies to be farmed as food but soon escaping into the waterways, the American signal crayfish has thrived at the expense of the white-clawed crayfish, the UK's native species, which has been infected by a plague carried by the transatlantic incomers. There were always hundreds in the river. We'd collect fifty or sixty and put them in a big tank in my lab where I'd cleanse them of all the rubbish they'd been swimming in, feed them, and then cook them as part of a big gumbo – a thick American stew – with Mark imaginatively adding a bit of rabbit, partridge, or possibly even squirrel to the mix.

Safe to say the bowels of that building saw occurrences mirrored nowhere else on campus. On one occasion, dozens of baby baboon spiders escaped. A large, fast, long-legged spider, not dissimilar to a tarantula, this was the kind of thing that most people would run a mile from, and as the arachnids disappeared down the corridors I assume that's exactly what happened.

Another time, the department actually came to the rescue. When a kingsnake was discovered dumped in a bin in the student library – it's a university, these things happen – it started a new life in our collection and actually bred with one of my own. From being coldly dumped in a bin to heart-warming romance – I can't believe Disney never picked up on it.

Insects, though, were my main bag. I was, and remain, fascinated by them, collecting and cataloguing well into my twenties. I still have my pet stick insects set as specimens in a small case. In fact, I'm picking up a beetle in a case right now, a rose chafer, an intense metallic green, which I can see I caught in France on 20 July 1997.

To soften a dead beetle, I'd put it in hot water. One mug for dipping, another with coffee to keep me going. All good, except after a while it's hard to tell which cup is coffee and which is beetle juice. Pick the wrong one and you know about it. As does anyone who hears you coughing and spluttering for the next ten minutes. I loved to lose myself in the absolute detail of the hobby. Self-immersion, the idea of a pastime that's somehow deeply engrossing, is not so common in these digital days, but for a while everyone seemed to have their own version of it – collecting stamps, growing roses, or whatever – a window into a different world, a holiday for the brain. We seem to have lost those portals, same as we've become removed from the natural world and its rhythms. You might think an angler is mad to sit and stare at a float all day, but think of that incredible mental escape, the fleeting glimpse of life in the water, the turquoise glint of a kingfisher. In that moment, that's all that matters. As a kid, I used to build hides in woodland and sit for hours with a flask of tea, just watching and waiting. We've gone from a world of patience to one of immediate gratification, from watching raindrops sliding down the window on a long car journey to going mad at the lack of 4G. We need to recognise that change – not so much harking back to a bygone era as understanding what we need today.

At Coventry, I instigated a specialised insect lab and also enjoyed my first taste of teaching, those environmental science sessions an early indication that I could combine knowledge with communication. I enjoy watching people respond to information. Give people the ability to understand their surroundings and it's like injecting them with a magical power. Having said that, at the university, if I'd had a big night at the Students' Union the night before, I'd just put a video on. I had a particular favourite about headlice, enjoying how the students would start to scratch while they watched it!

In more recent times, I've done live lessons for Children's BBC. It made me think: maybe, rather than berating students over a paragraph in a PhD, members of grand academic bodies and societies would be better employed going round primary schools showing kids a snake or an Atlas moth. There's a place for academia but not if it talks only to itself. I know that eventually, for me, academia took the pure excitement of loving animals and turned it into just numbers. Yes, research could be fun, but essentially it was just stats. There was a distinct lack of colour. That went against the grain. If there's a subject I'm excited about, I want to share it. I can trace that right back to setting up my first fish tank. I put some pondweed behind a rock and was amazed by how this makeshift underwater scene came to life. In that moment, I shouted for my mum to come and see. Excitement is good. I love it now when one of my own children wants to show me what they've drawn. We need to encourage that emotion, not smother it with dryness.

With limited satisfaction in academia, there came a point as I sat in my basement workspace counting flies where I looked at myself and wondered whether I might be leaving myself open to regret. Surrounded by four walls, swathed in electric light, my only escape a slice of the roof where I'd grow a few plants, this couldn't be how the rest of my life would pan out, could it? As my dad used to say, 'You're a long time dead.'

I didn't want to be one of those blokes in the pub claiming they could have done this or that. 'If it hadn't been for my ankle, I could have played for Tottenham' – you know the sort of character. I'd always been the type who'd much rather stick their neck out and have a go than die wondering. I craved fresh air, a more wholesome way of life. I needed to take a risk and try something new, find some freedom. Hesitate now and who knows? It might never happen.

I knew it was possible to do things differently. I'd read John Seymour's *The Complete Book of Self-sufficiency*, still the absolute bible on the subject almost fifty years after it was first published. The book is more than advice, it's an ethos. John's eyes were open to the ills of mass food production and the wholesale rejection of traditional farming methods. With his adherence to sustainability, he foresaw precisely the environmental issues we're experiencing now. Immersing myself in John's book was like a bright neon signpost appearing at a time when I was questioning the direction my life was taking.

Some people might have bought an old VW camper and headed off around the world. My idea was a little more concrete. I wanted to leave suburbia behind and find a place where I could build my own self-sufficiency ideal, not just on a back garden or smallholding basis but as a viable long-term business, farming more like it used to be, working with the land, getting back to basics – using proven methodology to produce good-quality, fresh seasonal food.

My head was full of thoughts of Simon, a friend whose dad kept rare-breed sheep and cows. Passionate about preserving such animals, he had Ryeland sheep, one of the oldest English breeds, as well as Dexter cattle, the smallest native breed in the British Isles, and Gloucester cattle, one of the rarest. A decade or more earlier, I'd seen my first lamb born with Simon. Tucked into my coat on a cold winter's evening after school, I watched as a Ryeland ewe produced an incredible little lamb, immediately christened Buttons, which struggled to its feet and eventually started to suckle. That was the point my mind became locked into the role of domestic traditional breeds of livestock. It made me think about the link between farming for the benefit of wildlife and for the benefit of the environment. Right there and then something really kicked in with me, and the basement of Coventry University was when it became crystal clear, one of those classic sliding-doors moments. Had I not gone the way I did, I expect I would now be well settled in a museum, hovering over a butterfly collection. I'd have been fine with that, but I think there'd be something in me slightly unfulfilled. If you have a real yearning, you can't really dampen it down. The fire doesn't go out.

Again, when it came to my farming ideal, Pannington Hall Farm fitted the bill perfectly. In this part of the world, arable is king, and yet the land here wasn't fit for crops; it was much better suited to pigs.

Commercial pigs had dominated the market since the development of the Common Agricultural Policy in the late fifties. To boost production after the war, farmers were encouraged to churn out food. Rare breeds, which are slower growing and have smaller litters, were the ones to suffer. Under the banner of The Essex Pig Company, what I would welcome to the farm was the saddleback, an amalgamation of the Essex and Wessex pig, reflecting the parlous state of both. The fact that the saddleback is the traditional East Anglian pig breed was a bonus – I immediately had a marketing tool, essential as to move forward I would need to start making money without delay. Later, I would add other rare breeds to the collection, but for now the farm's condition meant that for a long time the day-to-day was less about porkers and more about make do and mend. Initially, me and two pals – Rick, who I'd met at Coventry University, and Asa, an ex-Marine turned security guard I'd come across at one of Jamie's book-signings (Jamie was a big thing by then) – lived first in tents and then caravans, slowly, bit by bit, doing the place up. Michaela also threw herself into the project, travelling over from London every weekend. She'd got used to me by then – I think she had early warning of what she was letting herself in for when on our first date I invited her to an aquarium.

With no running water on site, I was pleased to find a well and, in time-honoured tradition, used a bucket to bring sustenance up from below. If that sounds a little picture-perfect, an image from an old Ladybird book, then bear in mind that I also had to fit a septic tank – in a day. Why in a day? Michaela's mum was coming to stay. The alternative didn't bear thinking about. I could hardly meet her off the train with a trowel.

Another time, Simon came to stay with his very glamorous Brazilian girlfriend Luciana, now his wife. They slept on the sofa in mine and Michaela's caravan – along with our three dogs and a piglet called Jasper. As they left, Luciana turned to Simon and said, 'Let's never come back here again until they live in the house!' For us, though, this was our normal. That first summer, we used to wander around in the evening sun with a glass of wine listening to Norah Jones and thinking everything was just brilliant. Which it was – well, most of the time. It could have been all over before it had barely started. The biggest emergency we ever had came in the run-up to a fancy dress party. A bunch of friends had travelled over to help out on the farm for the day, rounded off with a barbecue and disco in the evening. Even for a very warm summer, the day in question was crazily hot. As early as 7 a.m., as I went to let out the chickens, I could feel real heat on my back. By the time our army of volunteers turned up a few hours later, the sun was nothing less than blazing. Even so, everyone set about their assigned tasks eagerly, clearing bracken, chopping trees, digging up nettles. There was a lot of detritus and so, as is usual on a farm, the easiest way to dispose of it was via a carefully laid fire. Unfortunately, however, the flames spread, aided by a persistent wind. You see and hear about wildfires and their ability to spread quickly, but nothing prepares you for the shock of seeing that speed in the flesh. The fire's charge was astonishing and our best efforts to hold it back with buckets of water were never going to be enough. We were fighting a losing battle, one which could potentially take the farm buildings with it, as well as being a danger to us, and so the fire brigade was called. As smoke filled the air, the neighbouring A14 and railway line were closed. We'd announced our arrival, but not quite in the way we'd imagined.

With multiple fire engines on site and army green goddesses on standby, the fire was eventually brought under control. By way of a thank you, we invited the firefighters along to the party – well deserved by everyone after the insane efforts of earlier.

The next day, I was up and about early to inspect the damage. Getting up after a big night was doable then. When you're young you can drink a lake and be out and about in the morning. These days I only need to look at a glass of whiskey for it to hit me like a train. The irony is that the fire had worked wonders in burning away a layer of rough vegetation which would have taken me weeks by hand. I was able to turn that land into pasture much faster than I would otherwise. Within days, from what looked like a moonscape, amazing shoots of new grass emerged.

With so much to do, the fire soon became history, but it did show us very early on how easy it is for something to go dreadfully wrong, how important it is to be prepared for difficult situations, be it an accident on site or a storm bringing down the power lines. You can never second-guess the kind of emergency you might one day be dealing with.

•

Cutting back overgrowth, hammering in fences and shoring up battered old outbuildings were just a handful of an endless avalanche of jobs throughout those scorching months of 2003. After a day building pig shelters in incredible heat – the mercury reached 38.5°C (101.3°F) just a few miles away in Kent – a shower, which was singularly lacking on site, wasn't so much a luxury as a necessity. After some thought, I came up with an answer. In the morning, I'd fill some large containers from the well and then leave them in the hot sun all day long. I'd also found an old one-tonne bag – the sort of thing in which builders deliver sand or gravel – and cut the bottom out (for some reason, this particular shower cubicle never caught on). Come the evening, I'd pour the water over my head, soap every-where, and wash the whole day away. The transformation was so instantaneous, so massive, like putting on a new skin, that I can compare it only to Popeye after eating a tin of spinach. My own restorative tended to be a few beers and some much-deserved dinner. It was all so basic, and yet that was precisely what made it so massively rewarding.

Eventually we were ready to welcome the pigs – which was, after all, why I was here. A boar and eight females duly arrived. The day they turned up was incredible, a moment we'd been working towards for five months, but our porcine pals weren't interested in such high emotion. Instead, they went straight into land-clearance mode. As I built up numbers, I saw more and more just what machines pigs are when it comes to turning ground over. For too long, the only thing that had turned this land over was rabbits. Hens, ducks, geese and sheep were also perfect for clearing the weeds, which meant I could put more grass seed down and create better pasture, leading eventually to keeping cows. I know – it almost sounds like there was a plan!

Every day was a school day; we were always learning some-thing new. Many farms are passed down through generations. The knowledge is always there. Here, we learned as we went along, the hard way sometimes. Water supply is a case in point. Having taps at regular points on a pipe network across a farm is vital so you can easily turn the water supply off when you're doing certain jobs, like adding new pipes or fixing troughs. Being sprayed with water is highly amusing for anyone watch-ing but not quite so funny if it's freezing cold and you're the one being drenched. Even when you get the job done, saboteurs are lurking just around the corner. Many's the time I've proudly levelled a new trough, filled it with water, ballcock bobbing away nicely, only for a gang of pigs to turn up and start playing with it, nudging it around until it's totally off balance, water pouring out everywhere.

Aside from occasional pig-related tomfoolery and the gaudy pink interior of my caravan, those first few months were halcyon days because there was no major pressure. The fact I wore my inexperience on my sleeve, though, did mean people were forever trying to sell me stuff I didn't need. I had just enough about me to know that whatever strange-looking piece of machinery was placed before my eyes was unlikely to be hugely helpful in the keeping of rare-breed pigs. Other kindly people would offer me advice. Maurice was one of them. An old chap in a three-wheeled van, in which somehow he'd negotiate the pothole-ridden track down to the farm, he'd turn up unexpectedly with priceless tips on pig-keeping.

'What you've got to do,' he'd say to me, 'is give them a little bit of food but not too much, because a fat sow's no good, is it?'

It was Maurice who presented us with our first three chickens – 'I want to help you and Michaela get a start' – and would give us gilt-edged recommendations re: dumper trucks, not that we particularly wanted a dumper truck. Where he fell down a little was not being too au fait with women being part of a business. 'Buy your wife a frock every now and then,' he'd tell me. 'Take her out for the night. She'll get down and you'll want to bring her up again.' Actually, again, in his own way, he was making a very important point. Yes, the animals, the infrastructure, the equipment are all important, but what really brings strength to a farm is your relationship. That, right there, is the engine. Often on a farm you can be so busy that you simply forget why you're doing it and that actually your relationship is the absolute heart of it.

Simon's dad, who had so inspired me when as a kid I'd watched his Ryeland ewe give birth, gave us both financial and practical advice – 'Don't have six different breeds of cow, maybe just specialise in one or two.' Which I did – although I do have six now! Then there's Paul Kelly, whose dad, Derek, used to work for Bernard Matthews before turning his back on the mainstream and, at a time when white turkey was all the craze, made his name with bronze turkeys. Paul has been an incredible mentor, with an incredible knowledge of the intricacies and pitfalls of the farming business. It was so lovely that, from nowhere, and seeking no reward whatsoever, friendly faces would want to help. But I'm sure most people in the area thought I was nuts, and probably still do.

If we were going to recoup our initial expenditure, we would need to start selling our produce – and quickly. A shop was an absolute must. Typically cobbled together, I built a square box out of old fridge panels, while my brother bricked up the front so it looked like a rustic old farm outlet. To celebrate its completion, a gang of us piled into the pitch black of our newly acquired refrigerated van and headed to the pub. Asa was just getting into a tune on his guitar as the door slammed shut – which is when we realised the fan was on. The temperature plummeted to 4°C (39.2°F) and we emerged fifteen minutes later looking like semi-frozen zombies.

That van was another vital purchase. One of the few positives to come out of the foot-and-mouth outbreak of 2001 was an awakening of curiosity about food and its provenance, which itself led to a surge of interest in farmers' markets and the growth of farm shops. Many rural people realised that for the sake of their livelihood they were going to have to reinvent themselves. Processing and selling their own products was an obvious answer. The more you self-process, the higher the profit margin. For the customer, meanwhile, there's another benefit – more visibility. The horsemeat lasagne scandal came about because of an overwhelming lack of transparency. The further we remove ourselves from our food, the nearer we come to shock and horror. One of the most desperate sights I have ever seen was in California. I was filming a TV show when I first smelled and then witnessed a huge feedlot compound within which tens of thousands of cattle had been penned in a ploy to add weight quickly before slaughter – cows that roam put on weight more gradually. I looked over and saw these animals, barely an inch between them, standing on bare brown soil, not a blade of grass to eat. Feedlot cows tend to be given grain, despite the fact it can cause internal pain and damage their health. It was truly horrifying, a living, breathing example of just what we've reduced our food production to.

Thankfully, more and more people are demanding evidence that the food they eat has been responsibly and ethically produced. We have built on that desire for knowledge of the production process by actually showing visitors to the farm, young and old, how our meat travels from field to fork. We show them sausage-making in all its glory, something that seems never to fail to amuse. Indeed, viewers of my Channel 4 show *Jamie and Jimmy's Friday Night Feast*, with my old Blue Stratos-wearing pal Jamie Oliver, will know that I'm a keen proponent of the DIY pork sausage that anyone can make at home. I always like to remind people that we're part of the food chain like every other living organism, and processed food is far from a natural link in that chain.

We understood that selling our products at farmers' markets would be a great and reliable source of income. In fact, it wouldn't be exaggerating too much to say farmers' markets soon became our second homes as we pushed our pork sausages, a product that, with its combination of rare breed meat, herbs and spices, we felt would tempt people to pay that little bit more than they were used to paying for the somewhat blander fare in the supermarket.

At one point, we were doing six farmers' markets a week – hard work and yet at the same time liberating, emphasising again the idea of being out of the nine-to-five. I'd turn to Michaela and say, 'Isn't this great? We can do anything we want.' Maybe I was over-romanticising a town square in pouring rain just a little. When I think of the work that went into just one trip – curing the bacon, making the sausages, packing, unpacking, driving there, driving back, ticking stuff in, ticking stuff out, getting ready for the next one, etc., etc. – to return home barely having turned a profit, which could happen on occasion, could be a little demoralising.

Back at the farm, there was also the not-inconsiderable task of building the infrastructure, paddocks and the like, to give us the food to sell. Not to mention the actual day-to-day rearing of the pigs and taking them to slaughter. When you have an endless list of tasks, in your mind each and every one becomes more and more time-consuming. There were times when our brains were totally mashed. On one occasion, we arrived at Hadleigh, in Suffolk, baffled as to why we couldn't find the farmers' market we were meant to be attending. Eventually we pulled up and asked a pedestrian. 'I don't know what you're talking about,' he told us. 'We've never had a farmers' market.' There is, of course, another Hadleigh – in Essex.

And yet, on even the toughest days, there always seemed to be a ray of brightness. One elderly woman told us our sausages were the best she'd tasted since the war. Other times a word of encouragement, an appreciation of what we were doing to further not just the survival but the blossoming of rare breeds, would fuel us through a rough afternoon. I'll tell you now, any trader hopping from foot to foot in a big coat, woolly hat and fingerless gloves in the middle of November takes heart from anyone offering a positive word. Back in the invisibility of a muddy field or draughty workshop, it might be the only thing that keeps them going.

Certainly, I never lost that love of the markets. Heading here, there and everywhere in that little van, we felt like a travelling band of landlocked pirates. Michaela and I spent one Valentine's Day at Norwich farmers' market, selling sausages in the rain in a car park outside Sainsbury's. I bought her a rose and remember saying to her, 'What the bloody hell are we doing here?' But actually we were doing exactly what we wanted, and needed, to do, making money for the farm. For this special date on the calendar, I came up with an aphrodisiac 'love sausage' containing ginseng, garlic and Champagne. The idea was I'd sell it by the inch. Who would leave with a 4-inch love sausage when they could have a 6-inch one?

Another time Asa and I told two nurses they could have a free packet of sausages if they looked after the stall for five minutes. An hour later, we returned from a very pleasant trip to the pub.

Clearly, the bigger the event, the better the yield. We headed to the massive Ally Pally farmers' market in north London only to realise – Hang on! – we'd forgotten the small matter of the marquee. All we had was a fridge. Other days at 'the Palace' went slightly better. We'd sell our fare and also, donning my apron, offer hot food. When we made £1,800 in a day, it felt like winning the lottery.

Once we'd cleared out the barn, we began to stage our own farmers' market – stalls, up to forty of them, and straw every-where, with the odd animal thrown in. There'd be a sow and piglets in one space and then in the next there'd be visitors having coffee, while above, birds flew through a hole in the roof. We put a big blackboard up on one wall for kids to draw on – it was actually to cover the smashed bricks behind it, but who was to know? It felt like out of nowhere this lovely old barn had gone from being forgotten about – let down, considering what it had contributed over the years – to being an epicentre of life. Same with the old dairy yard, which we set up as a barbecue area. Again, nothing fancy. Just straw bales for people to sit on and a simple bit of kit from B&Q.

We were constantly looking for ways to get our products and name out there. We went up in the world by attending the

Food Lovers' Fair at Covent Garden, run by Henrietta Green, the famed writer and champion of local food. Henrietta settles for nothing but the highest standards and I was quite literally shaking as she checked our eligibility to attend. Even after she gave the green light, my nerves were jangling. Traders really go to town on the presentation of their stands – we had a pop-up marquee and second-hand chiller. It seemed Jimmy's Farm attending the Food Lovers' Fair at Covent Garden wouldn't be dissimilar to a battered old Skoda being parked on the forecourt of an Aston Martin dealer. Prior to the event I went to the supermarket to fill up the van with petrol and had a real stroke of luck. The hedges in the car park were rosemary. People were actually buying rosemary from the store having just parked their car a foot away from the stuff. Well, that was it. I headed straight back to the farm, fetched my shears and hi-vis vest, and got down to a serious bit of chopping and cutting, whistling a few songs as I did so to convince passers-by I was just a handyman at work. I chucked a whole load of the stuff in the back of the pick-up. The difference it made was incredible. Dressed in sprigs of rosemary, our stall looked – and smelled – magnificent.

We did another huge food fair at the NEC in Birmingham, notable for the amount of swapping that went on between traders. In exchange for some of our sausages, we came back with a chopping block, a set of saucepans and some beautiful venison. That was the thing with the markets, they had that real spur-of-the-moment element about them, the idea that to be a bit unorthodox is good. I loved that attitude and would like to think I've sprinkled quite a lot of it across the running of the farm and wildlife park. In fact, there's every possibility it was that very attitude which made that first series of *Jimmy's Farm* such a winner.

To be honest, no one was more surprised than me that my battle to bring the farm back to life with the assistance of rare-breed pigs should be deemed a winning format for prime-time TV. I know *Emmerdale* was doing well in the ratings at the time, but even so!

To be honest, *Jimmy's Farm* wasn't the first time I'd been approached about appearing on the box. Not long before, a director friend had asked me to do a pilot for a show going by the charming name of *Killer Diseases*, because for part of my degree I'd done a bit of medical and veterinary parasitology.

'So can you talk about tapeworms and stuff?'

'Er, well, yeah, I suppose.'

The show came to nothing and so my on-camera relation-ship with tapeworms began and ended there. But a few months later the BBC contacted me to see what else I was doing. When they heard I was starting a farm from scratch with zero experi-ence and money, and the idea was tied up with preserving rare breeds, they saw the potential for a documentary. They were, I suppose, on to a winner either way. The venture was destined for either improbable success or horrible failure. There was never going to be a halfway. An assistant producer was sent down to see what was what, *Jimmy's Farm* was commissioned, and the rest, as they say, is history.

We were still living in a caravan on site when the first episode of *Jimmy's Farm* aired. I had no idea what to expect. About half an hour after it ended there was a knock on the door. It was pitch-black outside. Tentatively, I opened up. There was a man, a complete stranger, standing there.

'Hello. Just watched the show. Thought I'd pop up and say hi!'

'Oh, OK. Thanks.'

'Anyway, all the best. Cheerio!'

No sooner had he arrived than he'd gone. A little bizarre, but then again I had no idea how anyone would react. All this was as new to me as it was to everyone else.

The next day, we went to the cash and carry, where after a few nods, smiles and glances, it dawned on me that more than just our late-night visitor had tuned in – very lovely and at the same time quite disconcerting.

As Jeremy Clarkson would find twenty years later with his own farming show, when the lanes of the Cotswolds were clogged as people flocked to his shop, TV does tend to change the situation. Our own shop got really busy really quickly, which meant a very steep learning curve. We'd have a lot of footfall and so we'd hire staff and buy extra bits of kit. Then, at the end of the month, we'd do the sums and realise that, despite turning over a decent amount of money, we'd somehow lost several thousand pounds. Clarkson's famed Lamborghini tractor was never going to be in our price range.

While that first brush with TV undoubtedly gave us a massive publicity boost, I was always conscious that the farm was never a made-for-TV project. The fundamental reason behind it was the preservation of rare breeds and the production of food. The TV element, welcome as it was, had happened almost by accident. I also didn't let it go to my head. I was on a self-imposed salary of under a pound an hour, while my clothes looked like hand-me-downs from Stig of the Dump. For most of that time I was also living in a mobile box on wheels with an old yellow settee, three dogs and a pig. Eventually, it was turned into a chicken house. To this day, I'm surprised they agreed to move in.

I was no stranger to unorthodox accommodation – for a while I'd lived above an undertaker in Walthamstow, a flat with no curtains, always interesting when a double-decker bus went past – but eventually Michaela and I made a move to a little cottage up the road, nice because it gave us a little bit of headspace away from the running of the farm. There was always someone on site but it gave us a release from the madness. There was another advantage. The Stour estuary was just over the fields. I'd heat a terracotta tile on a fire and on it cook freshly caught sea bass. There is a lazy version of this dish – wrap it in foil and chuck it in the oven.

It was only natural that customers at the shop should be inquisitive about the pigs that provided the meat they were buying. It wasn't so much a deliberate decision as one of evolution, but it felt like one minute we were a farm and the next we had people following a makeshift trail around the nearby paddocks. Well, might as well give them a bit of variety – so in came other rare breeds of poultry, sheep and cattle. Chuck in goats, ferrets (naturally), rabbits and guinea pigs, and families couldn't get enough and kept coming back. Right there really was the making of what we are now. Those first few paths were the precursors of the several hundred yards of trails, through the pasture, garden, orchard and woodland we have now.

There were always aspirations – gravel instead of hardcore, garden instead of scrub – but even so I look at old pictures from that time and can't believe what I'm seeing. That decrepit old barn, so forlorn, is now a restaurant. To have not just a shop but an eatery where visitors can enjoy the produce of the farm means a much more stable operation. And yet the barn very recognisably remains what it was. The beams are the originals, so intricate, delicate almost in places, but which have held the building up through all weathers for year after year. Straps from the timbers have been turned into door handles, while a lot of the metalwork has been fashioned from the old conveyor for the grain. We filled this incredible space with furniture made in England, all placed on wooden floors upcycled from an old factory. On the walls we have two sets of antlers. One is from a fallow deer and the other from a Formosan sika deer, endemic to Taiwan but now extinct in the wild. Both used to live at Mole Hall. I adore that I can look up and see that link with my past, that so much of what we have here has been created directly from existing materials. When we arrived, the ground was littered with old bricks. The obvious thing to do was hire a digger, drag them all from the ground, and sling them in a skip. But there is character in old brick. Why throw something out when it can still shine? Those bricks now form the walls round the garden near the entrance to the site. There is nothing that could look more natural, more of the land.

While farmers' markets had a lovely community feel about them, both in terms of the shoppers and the traders, persuading people to come to us made so much more sense. We gave up the big food fairs too. All the processing, packing and transporting made the cost prohibitive. Again, our view was that we could put on special events here, which is exactly what we have done. Down the years we've had everything from our own sausage and beer festival to a cockney knees-up sing-along and 'Big Beef' nights – a whole side of beef hung up in the restaurant, with the audience talked through how each cut is used and the ethics of how the meat is produced, while taking on board the small matter of six courses. I even do my British cow impersonations! Informative, entertaining and very tasty, all at the same time.

It would be tempting to think that after twenty years the waters of daily life are now millpond smooth. However, when every day involves the welfare of several hundred animals, and the only certainty is that there is no certainty, then there's always a chance of getting your socks wet, potentially even drowning. There have been so many times when what we're doing here hasn't felt like a viable long-term option, days when it felt like we were veering on the hopeless. I remember one winter I had a chest infection and feeling really ill, I was lugging straw bales around in the snow with a wheelbarrow. I came in, looked at the bank balance, as ever in the red, and had to start that awful ring-round familiar to anyone with a business who's needed to buy time – 'Sorry, but is it OK if we pay you next week?' That hand-to-mouth existence is tough. We couldn't borrow money against the property because we didn't own the farm. In the early days, that's why the farmers' markets were so important. Truth is it takes most businesses a good ten to fifteen years to turn a corner, to start making money that can be reinvested. Which is why it always helps if you throw yourself into something that's a passion not a chore.

Running a farm and wildlife park is a complex job. To do it you need knowledge that varies from animal biology to how to get an ancient tractor running after a harsh frost. If there's a more diverse job, I'd like to hear about it. But the massive upside of all that hard work is simple. At all times you're embedded in the landscape and the lives of those incredible animals that depend on it – a beautiful natural gift which you are forever unwrapping.

How that gift presents itself is dependent on something entirely out of our control – the seasons. Many have passed since I came here but I can honestly say each and every one, emotionally – and on occasions, physically! – has left its mark. At the end of any season, for better or worse, I will never be quite the same person I was at the start, and for that I'm grateful. No matter what happens, to so closely witness the changing of the seasons, to understand how they shape us, our animals and the natural world, will always be the ultimate reward.

THE SEASONS

These acres have gone through unimaginable change in the past two decades. The seasons have been the single constant throughout. This small slice of East Anglia and its ever-expanding group of inhabitants has seen springs late and early, summers of drought and downpour, autumns variously crisp and muddy, and winters veering from quick and mild to long, soggy and ice-laden. Genuinely, I have loved them all.

We are lucky to live in a country where the seasons are so markedly different and yet so inextricably linked, each, in its own way, absolutely and jaw-droppingly stunning, a vast and occasionally creaking cog in the life cycle. Together the seasons dictate plant, animal and human behaviour. It is the seasons that decide when animals are born, whether nature proceeds or stalls, the periods when we can work and those when doing anything is virtually impossible. For all the planning, the development, the improvements – a macaque hammock here, an escape-proof meerkat enclosure there – in their timeless hands the seasons hold the power.

That fact was instilled in me early on. In our first year, we did a lot of the fencing on the most beautiful spring days, the ground still soft from the wet and dingy winter that had come before. If there is such a thing as perfect fencing weather, the seasons had combined benevolently to provide it just then. If only we could have matched that perfection. Look carefully and you'll see the long line of stock fencing that runs down to the railway track is actually upside down!

You see, the seasons – not ourselves, a bank, the animals, or whatever – have always been our ultimate masters in this venture. Sometimes the accelerator, sometimes the anchor, ultimately they have shaped everything. That could have been a little disconcerting – the seasons are, after all, nothing if not unpredictable – but actually I have always felt totally reassured. From the very beginning, I wanted nothing more than to have that oneness with the year, to have country life stripped back to its forgotten basics, to gather the knowledge of previous generations who deeply understood and connected with change, nature and the environment.

Human instinct in recent times has been to battle against nature – to rein it in, bypass it, or simply bulldoze it, chemically or physically. I never wanted our land to feel in any way unnatural, both in how it looks and in terms of the animals that live on it. I knew that to make that happen I would have to work closely with the seasons and the opportunities they gave us.

Our wildlife park is four miles outside Ipswich, not on the African plains. And though it has thirty acres of woodland, gently gurgling streams and some rather lovely ponds, neither could it ever be mistaken for the Amazonian rainforest. I'm fine with that. What this land does deliver is an abundance of habitats to suit the wide array of domestic and non-native animals that have arrived down the years. We have more than eighty different species on site and never have we put a square peg in a round hole. Putting aside the occasional shocked passenger on the Norwich to London express, doing a double-take at a zebra as they flash past, I feel the wildlife park is a place where the animals are genuinely at one with their surroundings and how they change throughout the year. Those elements are wholly interconnected, and that's exactly how I see the seasons. To me, the sting of the coldest winter wind is in direct relation to the glory of the warmest summer breeze. The apparent deadness of the autumn mulch is a reminder of the spring bounty from which it is made and which it will soon once again feed. It's precisely that cycle that brings me so much joy and comfort and dictates every second of life here. Yes, there's the occasional kick somewhere painful (both real and metaphorical!), but it only takes one amazing vision – a sunset, a birth, a tree ablaze with autumn colour – for every blow to be worth it.

I love watching the flow of the seasons. When you're that close to them, you can sense their subtle movements and how they affect the animals day by day. Change, though, is undoubtedly afoot. In just two decades at Jimmy's Farm we have seen a marked shift in the climate – summers that veer between excessive heat and non-stop rain, autumns that seem to go on for ever, winters that disappear into an early spring only to snap back with a vengeance. No one can deny the world is at a tipping point, and if we're not careful, we can lose ourselves in the desperateness of the situation, feel as if there's nothing we as individuals can possibly do that would make even the tiniest difference. Rather than put our heads in our hands, though, we have to believe in the value of starting somewhere, of doing something positive, looking at that glass as half full.

While recognising that one person can't do everything, 'What can I do?' should always be the question we ask ourselves. None of us can solve all the Earth's problems, but we can act in our own sphere of influence. One of the ways to do so is to embrace nature. That's the same whatever our surroundings, be it a 140-acre slice of East Anglia or a flat in the middle of a city. As part of this book, I'll reveal a few easy ways we can reconnect with the seasons in our daily lives, be it by introducing a plant beloved by bees to a window ledge, adding a tiny pond to a garden, or even growing fresh produce from the contents of our fruit bowls and salad drawers.

I'll also show how we can use leftovers and food waste – 15 million tonnes of edible food is binned every year in the UK – along with the natural bounty around us, to help the planet, while conjuring up delicious and quirky treats that would otherwise pass us by.

I hope you enjoy both them and this journey through a year in the beating heart of my world.

Spring

A bringer of hope, a carrier of potential,
a welcome smile after the toughness of winter.
The excitement of the starting line – a race that will
go on for a year, before it all begins again.

Recently, I got talking to an old man in the woods. I never fail to be amazed by the immense knowledge of people who have spent their lives in the countryside – and this particular chap was no different.

'Put a stethoscope against a birch tree in spring,' he told me, 'and you'll hear it gurgling as the sap begins to rise.' I didn't have a stethoscope on me but knew straight away it was something I'd have to try, possibly in full white medical coat to puzzle passers-by.

In some ways, the revelation didn't surprise me. I'd tried some silver birch wine with a forager once.

'It's pretty powerful stuff,' he announced. He wasn't wrong. I'm not saying it was strong, but you know in old cartoons when a character swallows something a little fiery and you hear the sound of a train whistle and then steam comes out of their ears? It was like that.

'It's the lifeblood of spring,' continued my forager friend. 'Every time I've done this, I've always had kids straight off.' I could well imagine. Silver birch wine was much favoured by Prince Albert. He and Queen Victoria had nine children.

Already a father of four, I made a mental note not to make it a regular tipple.

It's not just the silver birch. There's an abundance of energy all around in springtime – an explosion after that long, cold winter. In the woodland, I'm always taken aback by the wild garlic that emerges, a carpet of white with the most amazing smell. Garlic becomes the default ingredient in our house for those few weeks – wild garlic vinegar, wild garlic olive oil, wild garlic in sandwiches and sausages. I've even had a nibble on it while I've been out and about working. Stand well back should you ever talk to me at that time of year.

That same vibrant flourish, that vigour, extends to the daffodils and then the bluebells, the monochrome filter of winter wiped clear by an intense burst of yellow and lilac, as if someone has gone mad with the colour button on an old TV. The leaves are super-green and, because the woodland hasn't yet donned its full canopy, a brilliant light shoots through the branches. Woodland that had been left naked, dark and skeletal, has found the key to its wardrobe and is in the process of reaching for its finest clothes. It's not alone. In and out of those flowers hum the bumblebees in their thick, shaggy striped coats. Queens, fresh from hibernation amidst the leaves or in the soil, scan for an upturned tree trunk in which to start a colony.

Early spring will always hold a special place in my and Michaela's hearts because that was when we first arrived on

WILD GARLIC OIL

Makes 450ml

In springtime, wild garlic is in such abundance. You only need to take in its wonderful aroma on a walk to realise its presence. We've so much wild garlic here that we've often been lost as to how to make the best use of it – although we've had great fun trying to work that particular 'problem' out.

Early on, a little bit of experimentation led us to discover the beauty of home-made wild garlic oil, ideal for brushing on pizza dough or adding to pasta dishes.

Ingredients

6 leaves of wild garlic
4 wild garlic flowers
3 sprigs of rosemary
4 sprigs of thyme
1 or 2 dried chillies
6 black peppercorns
450ml olive oil

Method

Wash the garlic leaves and flowers, ensuring there's no debris on them, and pat dry before dropping them into a suitable bottle – one with a cork or some way of securing the stopper – with the herbs, the chillies and the peppercorns.

Fill the container with the olive oil, seal tightly and keep for at least two weeks before using.

the farm. It was the perfect time, one of wonder, when we could see our new surroundings at their best. Not only that, but with every single part of this life being so original to us, it felt like each day brought a surprise.

It's those beginnings that mark spring out as such a special and reinvigorating season. To share our own start with the nature that surrounded us was a unique experience. We'd wake not to the sound of the city but to birdsong erupting, that beautiful chorus. Although essentially what sounds so delicately, so intricately composed is actually our feathered pals pitching their territory and screaming and shouting for someone with whom to share it – 'Look at me! I'm hot! And I've got somewhere to live! Mate with me, not him!' Some bring a bit of coloration into the mix. Blue tits, for example, get their bright yellow breasts from the caterpillars they eat – caterpillars which themselves have taken their yellow hue from oak leaves. The female blue tit isn't daft. She'll pick the male with the brightest yellow breast because it demonstrates he's the best at collecting caterpillars. Before they know it, this flourish of what can be loosely termed 'courtship' has turned to collecting food and feeding the chicks.

Take time, meanwhile, to look down at what's beneath your feet and you'll see great swathes of invertebrates emerging from the soil, triggered to start their adult lives at a time that, crucially, creates a food reserve for nesting birds.

Elsewhere, signs of renewed activity are obvious. New shoots, chewed down, offer clues that mice are out and about. Badgers, foxes and their young are suddenly appearing too, as are the disorderly and disruptive stoat and weasel families we so often see here – in an ideal world we'd be able to issue ASBOs. Hedgehogs are a little more orderly, eagerly tucking into the free bounty after waking starving hungry from that long sleep.

One of my favourite spring sounds is the cockchafer, otherwise known as the maybug, and, in these parts at least, the billywitch. The largest scarab beetle in Britain, I hear them as they blunder through the bedroom window at night, their noisy flight explaining yet another nickname, the doodlebug. Cockchafers spend between three and five years developing underground and just six weeks in their adult form. In that time, though, they certainly make their mark, banging into windows and providing a sizeable feast for any bat or bird lucky enough to get hold of one.

I love also to see the moths coming out, as, naturally, do the predators that feed on them, like the pipistrelle bats keen to build their strength after that long hibernation. Bats tend to prefer the warmer end of the spring weather, the dusk smattered with their dark little forms flitting to and fro. If they emerge too early and the temperature then drops, reducing the food supply, they become torpid, retiring to their roosts and vastly reducing their heart rate and oxygen consumption until conditions become more favourable. Think about it – really we're not that different. We open the door one day and we feel a natural energy, a slight warmth on our face. That lethargic cold winter is at last being cast away. 'Right!' I always think. 'This is it! Kicking off the year!' And then a cold snap comes and I can't wait to get back indoors. Unfortunately, unlike a bat, for me torpor isn't an option – there's way too many jobs to be done!

Spring often has that sting. Given half a chance it will delight in luring you into a false sense of security. Winter has gone, the sun is smiling, teasing, tantalising, with hints as to what the summer might bring, and then suddenly – whoa! – the wind's howling, the rain's coming down sideways, and you step outside straight into an ice bath. Trying to get tasks done in driving rain, like laying concrete or putting fencing in, can be so miserable, but even then I find myself admiring spring's punchiness, its inherent changeability that gives it the potential to be either utterly amazing or totally destructive, to hurl a bucket of water in your face one minute and bathe you in sunshine the next – a combination with an after-effect that is totally mesmeric. Even when I'm soaked to the skin, sunlight through water droplets is an undeniably incredible sight. People love fairy tales and sci-fi stories for their fantasy imagery, but woodland after a quick cloudburst trumps them both – as if it's been showered in crystals – a kaleidoscope of nature. Look into the trees and suddenly you're aware of all the spiders' webs, previously invisible but now dusted with the finest mist – thousands of spiders living in just a few square metres. Secret worlds revealed.

Stepping out of the woodland, I'll make sure to stop and stare at the landscape around me. While an autumn light tends to settle, golden, the last few flames of the fire, spring sunshine is clear and sharp. There are days when, for mile after mile, I can see incredible detail, like an unseen hand has polished the lens, worked the focus, and replaced my eyes with telescopes. Truly breathtaking to have such beauty laid out in front of me. No wonder so many artists have travelled to this part of the world to try to capture it.

Rain or shine, we try to centre our lambing around the movable feast that is Easter, a vitally important time in our operation when potential visitors have a bit of free time, are looking to get out again after winter, and, hopefully, are heading towards our gates. Not only does lambing make for a memorable experience for the kids on school holidays, not to mention our own girls – feeding a lamb stays with a child for ever – but for practical reasons we want the young hitting the ground when the grass is coming up, lush and ready. For many people, the sight of newborn lambs frolicking in the fields is the surest sign that spring is here. We absolutely share that excitement, with the added element of the potential of those new animals, what they might mean for the farm.

While we're all used to seeing vet programmes on TV where exhausted farmers are up all night with straining ewes – and like any farmer, I've spent my fair share of time with a hand in and around the rear quarters of my stock – the good thing about the old traditional sheep breeds we keep is that they pretty much get on with it themselves. It's much more of a naturalistic process. With our Norfolk Horns, for example, and other more primitive breeds, more often than not you turn up and the lamb is already there. Similarly, because we don't run big commercial flocks – it's not like we've got a thousand ewes in a shed – we're not pushing for constant twins and triplets.

During Covid, when daily activities had to be distanced, to have sheep capable of giving birth on their own was a godsend – ever tried to stay two metres apart from your colleague when you're helping to birth a lamb? But that's not to say lambing is always a stroll in the park. With around thirty newcomers each year, problems inevitably arise. There have been occasions when, as happened recently with Shelley, our Southdown ewe, the smallest of the British breeds, a birth becomes unexpectedly complex and the lamb doesn't make it. In farming, you're expected to wear that bruise and carry on, but it's hard.

Even when a birth appears routine and successful, there can still be complications. We have a small number of Cameroon sheep, which look a little different to the average UK sheep, coated as they are in hair rather than wool. Well, would you want to be zipped permanently into a thick woollen coat in Africa? Less than a thousand Cameroon sheep are believed to exist so when Persephone, one of our ewes, had twins, naturally we were delighted. Until, that is, it became apparent that one lamb had been rejected. Actually, it wasn't a huge surprise – she'd done the same the previous year. In human terms, rejection might sound unnecessary and cruel, but in the natural world there is often a reason. Lambs are not always the same size, in which case the ewe may be hardwired to go for broke with the one most likely to survive. Thankfully, while Stevie, our general manager, doesn't, at least to the untrained eye, look too much like a sheep, the rejected lamb was prepared to suspend its disbelief. In fact, it did more than that. As Stevie dedicatedly nursed the lamb through those early days – and nights – wrapped in one of his children's coats, it became thoroughly convinced he had been awarded custody. 'Littlefoot', as he came to be known, would follow Stevie everywhere. Tempting as it is to have visions of the cutest pet in history, it does neither us nor the animal any favours to go down that route. It took some doing to persuade Littlefoot to let go of the thought of Stevie as parent of the year and best pal rolled into one, but eventually, while reconciliation with his mum was too much to ask, he was able to integrate naturally with our native British breeds, such as the Devon Closewool, which went from the most popular breed in the 1950s to a population of just a few thousand now – amazing how quickly a breed can slip away once new farming practices and intensive food-production techniques are introduced. All of a sudden a common feature of the countryside is wiped out, deemed irrelevant. Of course, breed preservation isn't massively helped if a ram breeds with the wrong ewe. We have two African breeds of sheep, the Somali (white with a black head) as well as the aforementioned Cameroon (chocolatey all over). Either the Somali ram was a little bit lazy or the Cameroon ram was very up for it, but for a while we always seemed to end up with the Somali girls breeding with the Cameroon male. We've taken precautions now to make sure he can't get in amongst them. If sheep can smile, then he's had his wiped right off his face.

•

The fun doesn't stop with birthing. Each year, visitors are entertained by the sight of us doing our best to catch our lambs out in the field, the reason being that new arrivals need docking to stop flystrike, where maggots collect in dung caught on the tail. The repercussions can be serious if those maggots then start to penetrate the flesh. Docking entails applying a latex ring which cuts off the blood supply until the tail is shed. It's one of those jobs where you really do wish you knew how to say, 'Sorry, pal, but this is for your own good,' in sheep language.

We are always looking to add to our rare-breed sheep, recently adding a Kerry Hill, one of the most attractive of all sheep, with its striking black-and-white face. But as well as the sheep, our goats are adding to their numbers. We have Guernseys, a rare breed with a beautiful golden colour and equally lovely nature, as well as the small and amiable African pygmy, whose acrobatics in leaping up to any available high point makes them one of our most popular entertainers. This is, of course, inherited behaviour. In their native Africa, they need to keep a lookout for predators. Also, the higher you go, the greater the abundance of leaves to eat. Boer goats, meanwhile, originally a South African breed, are friendly and will consume with relish every morsel of feed that comes their way. We also have a Bagot goat, notably skittish, more of which later.

With our cattle, we tend to like the calves born, as is traditional, in spring or autumn. We want mum to be eating the best grass, because then her milk will be the most nutritious. It's obvious that the fresh spring grass will be prime fodder, but the autumn combination of mild temperatures and healthy rainfall means the pasture enjoys a second flourish, giving calves a chance to put some good weight on before winter. Last year, our British Whites had autumn calves, with their mums using that last flush of grass as an important feeding booster.

Be it the sturdy Riggit Galloways, shaggily handsome Highlands, or tiny Dexters, the smallest European breed, the cattle always make for impressive viewing. At the heart of the action is Lucky, our Riggit Galloway bull. The breed dates back to the 1600s, the word 'riggit' being old Scottish and referring to a streak down the back – Riggit Galloways have a distinctive white stripe running down their spine. You can't really miss Lucky. He weighs 700kg, most of which appears to be pure muscle. No wonder he turns out some fantastic calves.

The Riggit Galloway is one of my favourite breeds, right up there with the British White, which dates back a century further and, with its black and occasionally red extremities of nose, feet and ears, was kept ornamentally for years. We also have an Aberdeen Angus bull called Bruce. Both he and Lucky will be here for a few years, but eventually, to prevent inbreeding, they'll move on and a couple of new chaps will come in. It's the same with the pigs – it's important to try different bloodlines.

Of course, you can always mix things up a bit with artificial insemination. These days you can go online, choose the bull, and buy the semen. Should you so desire it could even be a vintage, coming from a bull that died thirty years before.

By law, all calves have to be tagged to be given a pass-port. Without that documentation, they aren't allowed off the farm. It sounds onerous but actually there's solid food-safety reasoning behind it. Every cut of beef can be traced back to its point of sale, and the bloodline of the cow noted too. The process basically entails clipping a yellow tag to the ear, the cow version of having a piercing, minus the telling-off from the parents – the only person this particular mother is interested in tackling is the one with the tag applicator. Bovine mums are uncompromising in their instinct to protect. That's why two people should always be present for this operation if done out in the field – one to tag, the other to keep a lookout for irate mothers. Both ready to run. The alternative is to end up hoisted into a hedge.

By now, you might well have noticed that we've given a lot of our animals names. Sometimes we come up with them our-selves. Known for their cackles, the kookaburras, for example, are named after Karl and Susan, the long-running and highly vocal characters from Aussie soap *Neighbours*. Other times we run competitions, but often an animal comes to us with a name already attached. I wouldn't like it if someone suddenly started calling me Wilf or Sidney and so we keep it on.

It's surprising how, in your head, the addition of a name really does suddenly make an animal a much more real char-acter. Pigs in particular tend to have markedly individual personalities. We had one sow who was a really good mum, very protective of her piglets, but was prone to real mood swings. It felt sometimes like nothing I said or did could ever be right – she'd always take it the wrong way. In the end, I called her Bob Hoskins because, like the low-life cockney gangster characters he was prone to play, she could turn at any moment. It could confuse people that a female should be called Bob Hoskins but, when you think about it, it hardly matters – there's not a single part of naming a pig after a Hollywood film star that makes sense! Neither does naming them after a famed country and western singer, but it didn't stop us naming two Tamworth sows Tammy and Wynette.

Naturally, our reindeer are named after all things Christmas – Dolph, Mistletoe and Rowan. No prizes for guessing why the aforementioned bull is called Lucky.

It should be noted too that of our farm stock only our long-term breeders are bestowed names. It would be foolish to do so with those that are going to end up on the shelves of the shop. It's only going to encourage an unhelpfully deep emotional attachment. It's often said we shouldn't anthropomorphize any animal, make it human and project our ideas and emotions onto it, but the thing is, you can't help it. Like millions of others, I grew up watching Johnny Morris, host of the long-running kids' TV series *Animal Magic*, playing the role of a zookeeper, going round various enclosures and attributing voices to the animals. After seeing that, how can you not think of animals, wild and domestic, as having certain characters? Basil the ant-eater is the perfect example. He's got a really inquisitive nature and can get quite excited when he sees something new. For some reason, I always think if he had a voice he'd talk like he'd got a blocked-up nose, a bit like Kenneth Williams – 'What is it? What is it? Go on, tell me what it is!' – as his tongue shoots in and out.

I do talk to the animals. Don't call for medical help – it's a natural response to them being in front of me more than anything. Disappointingly for any particularly educated tapir or reindeer, my conversation is more general chitchat about their welfare than international affairs. Over the years, I've come to see that the pigs get the best out of me conversationally. I like to make myself known to them, chat, calm and reassure them. Chances are I'm getting just as much out of it as they are – assuming they are getting something out of it! Just so you know, I draw the line at talking to fish.

Connections don't always have to be verbal. The most poignant often come when an animal looks you in the eye. It happens the most with the pigs and cattle; them trying to work me out, and me doing vice versa. Both sides, I'm sure, wholly failing.

Piglets always look like they're smiling. That's just fact. One hard to avoid as they're born all year round. For pigs, there is no breeding season. They're at it all the time. Boars are our live-in lotharios, the Barry Whites of the park, roaming their love paddock. The common belief when a boar, ram or bull is released into a field of females is that the male chases after the females. In actual fact, what generally happens is that barely has the male set trotter or hoof on grass than the girls are careering in his direction. It must be an unimaginably huge ego trip! Basically, the animal version of being in Take That in the early nineties!

To achieve the best quality, I match boar with sow to produce the next generation of pigs guaranteed to maintain the high standards we and our customers have become used to. Boris, our pedigree Middle White boar, is always in the thick of the action. He gets the best of all worlds. Later in the year he goes out on loan, sharing his genetics with other Middle Whites, particularly hitting it off with Mabel at a farm down the road. I've always loved how breeders work as a community, acting in the best interests of the rare breed that so fascinates them. Of Mabel's litter, for instance, one female will come to us as a future breeding sow. Should you ever wish to go into pig breeding, by the way, there's one part of the process that's particularly easy to remember. After insemination, a sow will deliver in three months, three weeks and three days.

Sometimes spring feels like you can't turn your head without being faced with an act of reproduction. As unlikely as it might sound, tortoises seem to have non-stop sex. I always think that's what it must be like in a retirement home, minus the lettuce. But it's not all sex, sex, sex – *Jimmy's Farm* was on TV pre-watershed for a reason. Spring also marks a shift for some pigs, our Tamworth Crosses, for instance, to the porcine paradise of the woods. Though they originated in Ireland, Tamworths are so-called because Sir Robert Peel, as well as founding the modern police force, brought several over to his estate near the Staffordshire town. Tamworths' aptitude for foraging, equipped with long snouts for just that purpose, allied to incredible strength and stamina for covering terrain, made them unsuitable for modern mass food production techniques and so, sadly, numbers dipped substantially to the extent that just a few hundred breeding females remain.

Tamworths love to snuffle amidst woodland of oak and beech, a by-product of which is the incredible flavour of their meat. Seeing pigs in their natural habitat is hugely rewarding for me. The possibility of doing so was one of the main factors that swung me to take on the farm in the first place. When I realised it came with 30 acres of woodland – in which, I'll freely admit, I frequently got lost in the early days (I couldn't see the wood for the trees, quite literally) – all I could think was how perfect it would be for pigs. In the spring, they could feast on a never-ending buffet of wild garlic, mushrooms and beetle grubs, while in autumn, there'd be a carpet of acorns, sweet chestnuts and beechnuts.

Pannage, the practice of putting pigs into woodland, goes back hundreds of years and benefits not just the pig – and therefore its owner – but the natural ecosystem. Pigs are essentially four-legged ploughs, looking for nuts and turning the soil as they go. Opening up otherwise compacted ground releases nutrients and allows seeds to germinate. As unlikely as it might appear if you've ever seen a pig crashing through the undergrowth, they are just about the best woodland regenerators out there. Pigs went green before we'd even thought of it, clearing scrubland better than any machine or chemical. I think back to that first year and the tangled acres of overgrown land, abandoned for two decades, that faced me. My initial intake of pigs cleared thickets, bracken, weeds and nettles so quickly it was like none of it was ever really there.

At that point, I'd like to think I had a bit of influence with the pigs. It wasn't long-lasting. I'm not surprised. I often say I don't know why they bother with the likes of me. Their considered breeding puts them way up the social ladder. As we've added more and more, so I've been pushed further and further down the line. When, as patron of the Rare Breeds Survival Trust, which seeks to ensure we don't lose the diversity of our native livestock, Prince Charles came for a look round, I'm surprised they didn't shove me to the back of the queue.

Steadily building on the Saddlebacks, we've added Middle Whites, with their characteristic snub noses and large pointy ears, and so endangered they're even rarer than the giant panda, as well as the Large Black, Britain's only all-black breed of pig, and reputed to hail from two boatloads from China which docked in Cornwall and East Anglia in the eighteenth century.

The 300-year history of the Oxford Sandy and Black makes its rescue from near extinction as recent as twenty years ago even more vital, while the Gloucestershire Old Spot, with its big lop ears almost covering its face and its distinctive black spots – a Gloucestershire must have at least one spot to qualify for the title – is forever popular among visitors. Finally, there's the Mangalitsa, developed in Hungary in the mid nineteenth century and the only surviving breed of pig to have a thick

WILD MUSHROOMS ON BUTTERY TOAST

Serves 2

It's not often we think of making the humble mushroom the star of the show. Sometimes, though, it's not a bad idea to give it centre stage and see what it can do, especially since it's one of the most sustainable foods we can eat.

For a really lovely breakfast or simple lunch, often I look no further than mushrooms on toast. While I will occasionally pick wild mushrooms (as ever, take great care if you go down this route), you can just as adequately get that woodland taste via the plethora of different varieties on sale in the shops.

Often, I'll use a mixture of varieties, frying them in butter, heaping them on white buttery toast and revelling in the absolute joyous simplicity of this wonderfully satisfying dish.

Beans on toast, spaghetti on toast – how about giving the mushroom a go?

Ingredients

30g butter (plus a little bit more for the toast)
400g fresh mixed wild or farmed mushrooms
sea salt and freshly ground black pepper
4 slices of sourdough bread – treat yourself to some
    decent-sized chunks
1 clove of garlic

Method

Melt the butter in a frying pan, add the mushrooms, and fry on a medium heat for 5 to 6 minutes until they are soft. Season to taste with salt and pepper and allow to rest. Toast and butter your sourdough bread. Cut the garlic clove in half and rub it over the toasted sourdough in a sandpapering action – effectively you are making your own garlic bread, letting the essential oils seep into the toast. Generously heap the mushrooms onto your toast and enjoy.

woolly coat. The Lancashire pig had a similar hairstyle but unfortunately is now extinct. All in all, that's quite a select bunch. In a pig *Downton Abbey*, they'd be upstairs eating off the best silver, while I'd be downstairs washing their silk bed linen and making a stew.

Sometimes it's nice to get away from all the heavy breathing and grunting. A favourite escape is to our little greenhouse. Peer through the window and you may well find me just standing there holding a packet of seeds. There's a reason for that – that packet sells a dream. I look at the picture on the front and am totally carried away by the thought of what can be created from the miniscule shapes inside. At one point, I was growing so many different seeds – courgettes, tomatoes, carrots, all sorts – I had to rein myself in. I'd plant six different types of courgette and then – who'd have thought it? – have scores coming up all at once. Down the years, I've had fridges full of courgettes, so many I couldn't give them away. Even so, my childlike enthusiasm remains. I tear open the packet, plant the seeds, and straight away I'm urging them – 'Come on! Come on!' – to pop up. I really do need to do something about my patience, or rather lack of it, but it is such a great moment when they start to germinate.

From the moment I first set foot on the farm, I wanted to grow my own veg. It didn't make any sense to have all this land at my disposal only to fill up a basket each week at the supermarket. Not that, as ever, the creation of a vegetable patch didn't come without some hard work. Initially, the area was one big thicket. Luckily, I had the machinery – pigs, geese and sheep – to clear it. The geese and sheep got to work on the vegetation and the pigs turned it over. We then seeded it and soon had all kinds of salad vegetables plus hearty staples such as beans. Spring is the time to put the hard work in, summer the time to enjoy the results.

Opening the restaurant meant we could redefine 'fresh', offering diners food grown literally a few feet from their table. Not so much carrots and potatoes, which would be difficult to grow in the required bulk, but more specialised veg and herbs, anything that adds a bit of interest on the plate, which would be expensive for us to buy in.

Sadly, we're not the only ones who enjoy a taste of the veg garden's wares. Putting seedlings outside means a war of attrition with the many predators that fancy a mouthful. I've produced loads of fantastic healthy seedlings only to see the whole lot consumed by a very happy mouse. That particular incident was like being in a cartoon, a battle of wits as I tried to catch the little blighter, finding the hole where it lived, and eventually luring it out only for it to run rings around me. I'm not sure I'm cut out for mouse-catching. On the very few occasions where I have actually apprehended one, I've ended up holding it in my hand, peering into its eyes and feeling sorry for it. I can't help thinking how as a kid I would have so loved to hold this mouse, and yet here I am treating it like a sworn enemy when, like me, it's only trying to live its life. I'd never have made it in the pest control business. Rent-O-Save just hasn't got the right ring to it.

Even if a mouse thanked me for its stay of execution by giving me a break, the rooks and pigeons were more than happy to replace it as my tormentors. At times, it felt like the rooks in particular were laughing at me. They'd pull out the seedlings and just drop them on the ground to die. In the end, I actually began to see the defence of my crop as a military operation. If I was to win the battle, I'd need back-up. And so I made sure to have a reserve army of seedlings to take the place of the first wave if and when they got wiped out. Of course, that meant there were some years I had shedloads of plants which weren't needed. Fortunately, a nice potted seedling always makes a lovely little present, or two, or several dozen.

One thing I don't miss about spring is the punchiness of the early nettle. When you're weeding the veg patch and the stingers

GROWING WILDFLOWERS

Cowslip, cornflower, meadow buttercup, ox-eye daisy, plantain, red campion, wild carrot, even the common or garden dandelion – you don't need a meadow to bring the beauty of wildflowers into your life, a little slice of backyard or garden works just as well.

While it's tempting to think the best way to mimic wild growth is to scatter seeds everywhere, actually it's a better idea to go for straight rows. They say there's nothing straight in nature, but in this case a little interference is OK, the reason being that, when the plants come up, you can get in there easier to reduce the aggressive grasses and weeds that inevitably pop up around them.

Plant in freshly prepared soil, nice and crumbly, in spring – use a broom handle to fashion a straight groove – and come summer, it's not only you who will be rewarded for your efforts, the butterflies and bees will be more than delighted with a welcome burst of nectar too. What to you is a visual feast, is essentially a luxury buffet to insect life, somewhere for pollinators to refill and then move on to the next spot on their list.

The key to planning your garden is to see it as a mosaic. Each individual area fits into a much wider pattern. Even the tiniest space can be part of a bigger story.

are coming up – Ow! – it's more like a bite than a sting. I was given a false sense of security around nettles by my chef friend Gennaro Contaldo, who I joined to gather ingredients for his fantastic stinging nettle gnocchi. Gennaro's own passion for food came from endless hours as a child foraging with his father and grandfather. So long as I said, 'Hello, my lovely!' as I picked the nettles, he claimed, I'd be all right. I wasn't. I got stung again and again while the little green monsters didn't seem interested in him in the slightest. Seeing my pain and bemusement, he insisted I was doing it wrong. I noticed after a while that Gennaro's lack of pain had very little to do with verbally cajoling the nettles and more to do with the fact he was brushing each plant towards him as he picked it. That way he wasn't breaking the tiny glass-like needles that inject the venom. Fortunately, the delicious gnocchi he served up made my discomfort worthwhile.

Nettles are massively underrated as a foodstuff, delicious and a great source of nutrients. For any budding tailors out there, it's worth bearing in mind that their highly fibrous nature has even led them to be used for clothing. In the First World War, many German uniforms were made predominantly out of stinging nettles bashed into a fabric. More than any-thing, though, they're really important for many invertebrates, particularly butterflies. There's a whole group of nymphalids, the largest family of butterflies, such as the tortoiseshell and the peacock, that lay their eggs on stinging nettles, cater-pillars then forming web-like structures and moving as one from one leaf to another. We all want to help butterflies and often think flowers are the answer. They are, but that's only half the story. Without patches of stingers, we're not going to get more butterflies. Juvenile butterflies don't eat any flowers whatsoever. Humans like putting things in camps – good and bad. Nettles are a case in point, but there's a much wider story at play.

While it's unlikely nettle seeds will ever be flying out of the garden centres, there is definitely something wonderful about growing your own plants, how in the tiniest of greenhouses, pokiest of gardens, you can produce so much food. When I was a kid, my dad nurtured vegetables from seed, while my grandfather loved growing flowers such as roses and begonias, but beyond that it was something for the old boys on the allotments, competing at the village fete in categories such as 'best onion' or 'largest marrow'. Nowadays, growing your own is quite trendy, and rightly so, because there's nothing better to connect us with the seasons than planting seeds, or even a tree – the latter being something that so few of us ever do but which delivers such an incredible feeling, not only of renewal when it goes into the soil, but of being attached to the passing of time as we watch it develop. For all the digital technology available to children, there will always be something magical about planting something and seeing it flourish. And even in the tightest of spaces, it's easy to do. Did you know you can make a mini greenhouse from a plastic bottle?

It just so happened that the child me was more fascinated by animal than plant life. The food I was initially trying

New lambs are endlessly irresistible.

Carry a bucket and animals will follow you anywhere!

Prince Charles's visit was an incredibly special day.
He's passionate about rare breeds.

Thankfully, I managed to answer all his questions!

Gloucestershire Old Spot – free ranging is a wonderful thing.

Sebrights face off – 'Watch the feathers, pal.'

↑ 'You laughin' at me?' Kookaburras are essentially giant and unavoidably comical kingfishers.

↖ Alpaca – I had a similar haircut for a while.

← Close-up – the guanaco is a close relation of the llama.

↙ Redefining the word 'cute' – baby tapir, Tobias.

↓ Close-up wallaby – nature's use of texture and pattern never fails to amaze me.

to produce came in the form of eggs. I had three bantams, acquired from someone at school, and the first egg they laid I carried proudly back to the kitchen. Bantams produce tiny eggs, but that one, fried, and carefully cut into quarters, provided egg on toast – well, sort of – for four of us. Tucking into my portion, it didn't matter how small it was, the point was it was spring and my bantams were laying, and that was an amazing feeling. I allowed the hen to sit on the rest of the clutch and within no time she was parading round with a dozen little chicks. I had the best time watching them getting plumper and more bold throughout that spring and summer.

Again, hens are worth considering if you're looking to try your hand at producing your own food. They're basically a feathered recycling unit – feed them your garden waste and out comes an egg, a jewel, and again this is something through which we can track the seasons. As time travels from spring to autumn, so beta carotene levels in grass and foliage change. Beta carotene is one of the most common food colourants in nature, the reason why we see pink flamingos – it's present in the algae, shrimp and larvae that make up their diet. As the year goes on, so the egg yolk changes. When the beta carotene is rich, the yolk is a lovely vibrant yellow, perfect for the absolute delicacy that is the fried-egg sandwich. When that yolk oozes out the side and, let's face it, all over your fingers, it should be of the stickiest, yellowest kind. As the beta carotene starts to wane, so the egg becomes paler. The same goes for cheese. 'Turnout cheese', made from the milk of cows that have eaten early – turned out from their winter shelter onto the grass – is

GROWING SEEDS FOUND IN YOUR HOME

When it comes to planting seeds, we tend to think of buying packets from shops. Actually, chances are we have several just a few feet away in the fridge. That gooey old tomato could be the source of several dozen lovely ripe fresh ones.

It's not something that people often think of doing, possibly again because of the disconnect between food bought in supermarkets and the natural world, but it's something I've always done, and something that took on extra emphasis during the early months of lockdown, when growing food instead of buying it, thus avoiding a trip to the shops, was a topic of conversation. It was spring, and with so many places shut, it occurred to me that going out and buying seeds isn't always necessary. If you have a bit of potting compost to hand, you can take those old tomatoes, squeeze out the jelly, plant the seeds and then watch as they germinate and grow into beautiful plants.

It doesn't stop at tomatoes. You can do exactly the same with pumpkins, positively bursting with seeds, as well as dried chillies and peppers. Take a look in your cupboards. Chances are you'll have coriander seeds in there which perhaps you grind up for a curry. Put some of those in compost and soon you'll have a mini coriander plantation. That applies to fennel and cumin too. And fruit works exactly the same. I've successfully grown lemon and orange plants from the contents of the fruit bowl.

> The options are endless. You don't even need a garden
> – stick the seeds on a window ledge and you're away. Look
> at the food in your house with a different set of eyes. The
> source of your next bounteous crop might well be right in
> front of you.

often a dark, deep, almost orange colour. While an egg yolk's colour is dictated by environmental conditions, the colour of the shell, meanwhile, is dictated by the breed of chicken. A Maran hen will produce a dark chocolatey egg while a Leghorn will produce one that is pearly white.

The first hens we had on the farm were Warrens, low maintenance and prolific layers, as well as the highly companionable Light Sussex, thought to date back to Roman times. Soon came the Black Rocks (a particularly hardy Scottish breed), the Anconas, with a mix of black feathers and orange eyes, which can be traced back to the Italian city which gave the bird its name, and the Silkies, bundles of white fluff that really do look like they should be on sale in a toy shop.

We have also rescued birds from a caged system. We had a hundred at one stage. Factory farming turns an animal into an equation, at which point it's forgotten that this is actually a sentient biological organism. For us, it boiled down to this – those birds have been part of our food production system, let's give them something back, treat them with respect.

When those hens arrived they looked, as you'd imagine, like they'd had a hard life, the equivalent of someone who'd been put to work non-stop in a coalmine. They were raggedy, had feathers missing, and, in all honesty, some didn't last long. But, for the most part, their feathers came back in a couple of weeks and they were soon scratching around for food in the manner of a free-range chicken.

Batteries don't exist any more in the UK – nowadays, it's caged systems. The laying life of such birds tends to be about eighteen months, after which their eggs peter out and they're no longer considered cost-effective. While going for tradi-tional rare breeds is lovely, if you just want a few chucks for the backyard, it's worth getting in contact with hen rehoming charities. You won't get an egg a day out of what was a caged bird, but keep four or five and you'll get enough for the family. Chances are too that they'll be a little more interesting than shop-bought eggs. As with fruit and veg, supermarkets present only a neat and tidy version of reality. An egg deemed not to fit the perfect oval template will be discarded, whereas your eggs can perch on, wobble off, or sink into your egg cup as much as you like.

Among our rare breeds, we have Buff Orpingtons, a tremendous-looking bird which always reminds me of a rather overinflated posh person, puffed up with their own importance. The cock is the poultry version of the Wild West gunslinger – strutting, chest out, a pair of big old chaps on his legs.

Pick one up and you soon see that they're all fur coat and no knickers, like a big hairy dog that disappears to nothing the minute it gets in the bath.

The Light Sussex are a beautiful white with a black necklace, while our Silver Sebrights are tiny and, with their feathers laced in black, really do look like they've been hand-painted. We also have the Araucana chicken, named after the area of Chile from which it hails. The Araucana lays the most gorgeous blue eggs, one of those colours that only nature can create.

We keep our chickens in separate groups across the farm to prevent them interbreeding. We also make sure they're right there in the wildlife park alongside the capybaras, meerkats and tapirs. That way we can send out the clear and quite shocking message that there are features of Britain's natural world in danger of being wiped out too, something to think about if you ever decide to take the plunge and add a couple to your back garden. Maybe, by doing something as simple as keeping hens, you could be part of conservation and preservation too.

Whatever your plan, good luck in deciding which chicken to take on. So incredible is the variety that it's very easy to lose yourself in poultry breeds. I enjoy going to a country show and seeing so many gorgeous birds in all their different shapes and sizes, testament to how breeders have changed their form and function, and a far cry from those sorry specimens we see emerging from caged hen systems. Strange to think that each and every one of those various chicken breeds originate from the junglefowl, a wonderful little bird found in India and Nepal, where they scurry around the scrub of the forests. It's that small tropical ground-dweller we have to thank for our egg and soldiers and Sunday roast.

Unlike hens, which lay all year round, geese lay only in spring. They make up for that by producing eggs that are eight times as big. Nothing is lost in the flavour – goose eggs have a real intensity – you just need to make sure there's a few people around to help you eat one. Plus you need to factor in ten minutes hunting around for something – a jug perhaps, or a plant pot – that might double as an egg cup.

Early on, Michaela and I had a smattering of beautiful white geese on the farm. As time went on, we brought in greylags, the largest and bulkiest of all the native wild geese in the UK, and the ancestor of most domestic geese. We also have red-breasted geese, a bird under threat in its natural East European home but which is happy to make its presence known around the park, where its piercing call offers a lightning bolt up the spine to anyone within 50 metres – shocking, but well worth it if we can help with their conservation.

Throw in a few ducks, including the ridiculously pretty mandarin, with the male's incredible palette of oranges, greens, blues, purples and blacks, as well as the brilliantly named Khaki Campbells and a smattering of Indian Runners, with their comedic upright stance which meant they were often referred to as penguin ducks in the past, and it's easy to see just how poultry brings not just colour and variety but a huge amount of character to the farm and wildlife park. I remember how quiet the place could be when we first arrived. And then the ducks, geese and hens came along and suddenly it was so, so alive. The honking of the geese makes them sound really cranky. Of course, what they're really doing is self-preservation, using loud noise as a protective measure. When later we brought in guinea fowl, they too seemed never to stop making a noise, like a group of wittering old football supporters discussing their team's defensive frailties in the pub after a game.

At first on the farm we didn't realise how big a feature water was, but eventually we unearthed tiny streams running through the terrain as well as several bogs. Once we saw the water sources, we set about creating pools, an absolute mecca for wildlife, a place for birds to drink, nest and feed and insects such as dragonflies and damselflies to hunt, mate and lay eggs. Have a pond next to woodland or a meadow and it just erupts. Pond skaters seemed to arrive within seconds; water boatmen raced across the surface while dragonflies and damselflies flitted above. Newts, frogs and toads weren't far behind. If you want to increase biodiversity, a little pond in the garden makes a huge difference. Ours on the farm have certainly done so. As the weather warms up, they offer a beautiful oasis, including for the bees who rely on water for their honey production – it's a real treat to see them clustered on the edge of the wet mud taking their fill. In that first spring season, and those that followed, I would look up into the skies and wonder when the swallows and the swifts would arrive. And then one day there they'd be, swooping down to the water. Even now, I wonder if it's the first drink they've had since Africa. Obviously not. But in my head I can't help comparing their arrival pondside to that lovely relief of a cold glass of water on a hot day. In essence, the ponds are our version of the drinking holes found in the savanna; it's the same story of reinvigoration and survival but with a different set of characters.

This is nature, so ponds, as well as being havens for new life, can also be the spot for an early demise. A heron stalks the site and, as well as the indigenous fish we've added, such as tench, rudd and carp, ducklings are what it's after, spring being prime mealtime with numerous clutches being hatched. To a heron, goslings and ducklings are essentially a mobile buffet.

While a duckling being plucked from a pond by a giant spear-like beak isn't something that pops up in many bedtime stories, it is part of nature. As such, I accept it's going to happen. Where it becomes problematic is that amongst those being taken are birds being bred for conservation. The Muscovy duck, a native of Central and South America is a case in point. It tends to get ignored a little because it doesn't have the good looks of most other ducks – the male's face is covered in fleshy lumps – but I find them really interesting. Every year, though, the heron picks a few off. To be fair, I don't think it's wholly to blame. I'm pretty sure our pelicans – Peli-can and Peli-can't – tuck in too.

Herons are magnificent birds and normally I'd love nothing more than to have one on my doorstep to watch and admire. However, for the sake of our other pond-dwellers, I really did need this one to move on. The question was how to persuade it. In the wild, herons, like any animal, will do their very best to avoid predators. It occurred to me that we have a wolf. Not a real one, rather a plastic one set up in the woods as an example of the wildlife once native to these isles. I've done all sorts of weird and wonderful things in my time but wandering round a wildlife park with a plastic wolf under my arm has to be up there with the oddest. Pond expert Dolly and I placed the beast alongside the heron's favourite observation point. Our resident geese definitely weren't happy, heading straight for the opposite side of the enclosure. The heron? He couldn't have been less bothered. Same when I fetched the plastic crocodile from outside our tropical house and put it in the water. He couldn't have cared less. Somehow he knew this was just sorcery, none of it was real. There was one thing we could do to bamboozle our gangly-legged visitor, however – rear the ducklings inside. Once released, they'd be too big a challenge for the heron to risk trying its luck. For the same reason, we sometimes incubate eggs indoors. A bit late to protect the chick if the egg is taken from the nest! I challenge anyone not to be both moved and astonished at the sight of an egg hatching. It is a truly incredible sight. A tiny chick, so vulnerable, forcing its way through that shell into the outside world, utterly at the mercy of whatever it finds there, is nature at its most remarkable. Occasionally, if it's really struggling, or needs to be separated from other birds, a chick will come home with us to be nursed by the girls. There's not a child anywhere who wouldn't love that.

•

BUILDING A POND

We have everything from grass snakes to toads, frogs and newts in our ponds. The insect life is phenomenal and there are birds nesting in the reeds, including a family of moorhens. No matter how big or small, a pond makes a massive difference, and is, as you might expect for what is effectively a hole in the ground, really simple to make.

As a kid building ponds, I used to dig a shallow pit and line it with old bits of plastic from my dad's building materials. The plastic was never particularly easy to work with and inevitably didn't last long. Nowadays, eco-friendly pond liner is much easier to come by, but that doesn't mean you have to forego the idea of being a little different and doing it yourself. Old kitchen sinks, for instance, make great little ponds, but really any container will do the job so long as the sides aren't too steep – it's vital that wildlife can get in and out.

With any pond, I'll make a deeper area in the middle and then have a shelf where I can put plants. Not only do varying depths attract different animals, but shelves again make for a less hazardous experience for any passing hedgehogs taking an unintended dip.

In no time at all, animals will be coming to drink from the new addition to your garden, insects laying their eggs in the water, frogs and toads claiming an annual residency. Put a pond next to wildflowers or the edge of a few trees and you'll see an even greater abundance of wildlife as the areas naturally interconnect.

Don't forget to add some plant-life to the water itself; a mixture of submerged, creeping and floating plants form a vital part of the pond ecosystem. Dragonfly larvae, for instance, use plant stems to climb out of the water in the final leg of their journey into adulthood. It may well be that within days, a female is depositing her eggs back into the pond, the whole cycle starting all over again.

For us and for wildlife, the pond is the garden gift that never stops giving.

It's not just the tasty treats on the surface attracting the predators. Kingfishers and wild otters are always on the lookout for an easy meal. While the otters dismay Dolly – as a keen fisherman, he sees them as rivals on his patch – I can't get enough of them. For anyone who loves British wildlife, the presence of otters is like a lottery win; they're such sleek and elegant creatures, remarkable in their mastery of their environment (did you know, for instance, otters can close their ears underwater?). It's not just fish that make up their prey – ever-enterprising, a wily otter will also seize a chance to eat unobservant or vulnerable waterbirds.

A kingfisher, meanwhile, reminds me of my childhood. I'd hear its high-pitched whistle as it fired down the river, feathers shimmering electric blue like a shot from a laser gun. Many times I'd catch its call, turn my head, and already it would be gone. I love that the kingfisher, same as the robin, is always with us. It doesn't migrate. Thing is, I can't see a kingfisher these days without thinking of its Australian relations in our wildlife park – those great big chunky kookaburras.

There's one pond noise more than any other that just shouts, or croaks, 'Spring!' and that's the constant clamour of our amphibious friends trying to find partners, basically by jumping on anything within reach. Whenever I look at this mayhem, I'm reminded of the old days of Saturday lunchtime wrestling on ITV, except instead of Big Daddy battling away against Giant Haystacks, I'm watching a gnarled toad trying to get its grip on a mate while kicking a not inconsiderable number of love rivals away. This madness is soon translated into hundreds of tiny toadlets leaving the pond – not viewing for the faint-hearted. While they keep coming in waves, so do the predators – rooks, grass snakes and herons (naturally) included. Those that do make it to safety will hide up, feeding on tiny insects, eventually returning to the same pond in a couple of years to start breeding themselves. And so the whole cycle begins again. Remarkable creatures, although they can give you a heart attack when one jumps out of your watering can.

The same way as I was glad of the water, so I welcomed scrubby areas on rough or unkindly undulating ground. I'd watch in spring as bumblebees worked their way into the tall grass to create new colonies, or a kestrel hovered above seeking an unlucky mouse. Like I say, diversity isn't something to fear, it's a safety net – natural wealth. I was posed a question once. It came from someone who farmed on a massive scale in Brazil.

'What do you want,' he asked, 'food or flowers?' I felt to pitch food and flowers against each other – the idea you must have one or the other – was a big mistake. Truth is, you can't have one without the other. If we turn this planet into a factory floor, then we do so at our own peril. Natural systems don't work like that, they're not linear. If you think you can control the natural world by dictating just a few elements within it, then you are mistaken. All you have to do is peer into a grassy verge and see the umpteen species of grass, the hundreds of different invertebrates that make it tick. And yet we think we can manipulate the environment with just the odd twist of a knob here, the turn of a dial there. Most people have trouble remembering a phone number, so how we think we can regulate the world is beyond me.

The more diverse an environment, the more stable it is, the less likely it is to whack out of kilter. The more we erode diversity, the more unstable and less productive land becomes. Think about it: if, like many modern farms, you harvest a monocrop – growing the same thing year after year – then the law of nature says you're going to just attract one type of pest, and a load of it, which then has to be sprayed off, all of which affects long-term stability. Have lots of different organisms working alongside one another, preying on each other, and soon you have a robust environment, a natural piggy bank, putting nutrients back into the soil. For those who rely on the land, bogs and ponds, deemed 'useless' by some, help drainage and ease flooding, same as that 'wasteland' of scrub is full of fantastic allies.

It's exactly the same ethos that applies to rare breeds. They're not relics of yesteryear to be found in living museums. Adding to genetic diversity, they're as much about the future as the past. With the environment changing at speed, who knows what challenges are coming our way? Those rare-breed genetics might be just what save us when commercial breeds prove unable to manage in changeable conditions.

I always feel we can learn from history, not least its propensity to repeat itself. In the last twenty years, more than ever we have seen that big agriculture, with the damage it can inflict on the environment, is not the answer, and so smaller family farms, rather than being pictured only in dusty old oil paintings, have been reborn. It might sound like an oxymoron, but what we have here is essentially a modern traditional farm, one which, more than profitability and output, is about community. A farm used to be the centre of local life – people would go there for eggs, to buy poultry, fruit, vegetables, all sorts – rather than a massive food-producing operation far removed from the public it feeds. During the pandemic, with well-stocked shelves of fresh produce, farm shops reconfirmed themselves as a huge part of the local community. In our case, for example, we could offer farm-produced, rare-breed, free-range and traditionally butchered meat at a competitive price.

At times, we've had anything up to seven full-time butchers on site. As ever, I learned my own limited butchery skills on the hoof. I could make a very rudimentary job of a carcass but to be a good butcher takes a lot of training. It's all about minimising waste and carcass utilisation. Blunder in and you can very easily ruin hundreds of pounds' worth of cuts, so I concentrated on curing bacon and making sausages instead.

It was a revelation when we opened the farm shop. At that time, traditional high-street butchers were having a hard time, with many shoppers switching to supermarkets or the discount chain butchers. However, people loved what we were doing, that old authentic emphasis on quality, to the extent the stock was moving so quick I didn't have enough meat to fill the counters. One day, I had a thought – always a dangerous time for the business, but on this occasion it actually worked out. Often, a pig returns from the abattoir essentially split in half from head to toe. The shape of the pig is basically intact. I looked at those empty display cabinets and put two whole sides of pork in there. Immediately, people were fascinated. It was so far removed from the pre-packed and pristine uniformity of the supermarkets. I took great delight in both that and how much people wanted to learn about the huge diversity of cuts available to them. That's what our traditional high-street and farm-shop butchers give us. Go into a supermarket and ask for trotters and they'll think you're trying to buy drugs; go into a butcher's shop and you'll be much better served. I know also that some people feel a little intimidated ordering from a butcher because we've lost the knack of buying by weight, whereas supermarket meat just has a price on it. Anyone over a certain age can navigate a butcher's counter like an expert. But remember they're running a business – they're more than happy to help! I'm no stranger to butcher embarrassment, except mine is more Englishman-abroad based. I'll go into a butcher's on holiday in Spain and, foolishly, ask for different cuts in their language, getting progressively quieter as the ineptitude of my pronunciation becomes increasingly clear!

I know there are some who see the idea of a farm and butcher's as incompatible. I have heard others in the same business say that visitors wouldn't like the idea of seeing sheep, cattle and pigs in the field and then seeing their meat in the farm shop. But for us that cycle is at the core of what we do. It's really important we don't hide from it.

Honesty and openness about food is vital. It's like abattoirs being seen as places where terrible things happen, whereas the truth is that the vast majority are professionally run, an unsung and unseen linchpin of the livestock industry. The fact that we screen them off, never talk about them, because we don't want to face up to the reality of slaughter – out of sight is out of mind – is precisely what allows episodes of malpractice to take place. I get that a six-part series on the work of an abattoir is unlikely to be a ratings winner, but it can't be denied that it's part of the process, a cog in how we, as people, choose to feed.

I once travelled to a huge industrial abattoir in Denmark. What fascinated me was the transparency. I walked into reception and right there were pig carcasses hanging up in a big glass chiller the size of a football pitch. It was an abattoir with a visitor centre – imagine that! They had 120,000 kids come over every year. Initially, I was astonished. It was just so utterly unexpected. But then I had a thought – 'Hang on! This is exactly as it should be! Totally transparent. No hiding.'

Physically disconnect yourself from something and it soon becomes alien. The minute we wrap ourselves in cotton wool is when problems arise. In all honesty, I'd put food production on the national curriculum so pupils are given an authentic view of just what's involved in producing what they put in their mouths. Visiting farms, factories, even abattoirs could well be part of that. It honestly amazes me how little children are taught about the natural world, full stop. For whatever reason, the knowledge that millions had as a matter of course – be it knowing a bird from its song, a tree from its leaves, a creature from its tracks – has somehow permeated out of our culture, and yet the natural world is at the heart of our lives – what we eat, what we breathe – whether we realise it or not. Why should the essential fabric of our existence be so alien?

We're lucky to have built a great relationship with an abattoir down the years. The majority of abattoirs are contracted for large-scale meat producers, supplying supermarkets. They deal with commercial hybrids, very little variation. Abattoirs that kill just a few pigs or cows of different breeds, shapes and sizes are harder to come by. Pigs with coarse hair, for example, or of unorthodox sizes, present different challenges, and may be seen as time-consuming. Those who maintain and cherish the required specialisms are a vital part of rare-breed preservation.

The actual act of sending animals off to slaughter can never be a moment of empty functionality. At the start, I used to load the pigs up on the trailer and Michaela and I would say goodbye. They'd look straight back into our eyes. That's not

TOAD IN THE HOLE

Serves 4

Toad in the hole is one of those amazing treats born from out-and-out simplicity. No frills, no mucking around. Basically, all we're talking about here is sausages and pancake batter, a winning marriage between bangers and Yorkshire puddings. When those two worlds collide, something truly special happens. Add gravy and you're in heaven.

Forget the Hubble Telescope or the Large Hadron Collider – toad in the hole has to be up there vying for humanity's greatest creation. Spring, summer, autumn, winter – it doesn't matter – it's a winner. I've never seen any kid run home for a salad, but serve up toad in the hole and they're like Usain Bolt. It's on the table – bang! – and smells fantastic. Bite into it, and it's just everything – understated comforting familiarity.

For me, it also represents the world around us – the wheat in the fields, the free-range chicken eggs, the sausages from the rare-breed pigs. If it was French or Italian, it would be called something like 'farmers' village pie'. Being British, we call it toad in the hole, because that, apparently, is what a sausage wrapped in batter looks like. I often wonder what foreign tourists think when they see it written on a menu!

Ingredients

equal quantities of plain flour, milk and eggs

sea salt and freshly ground black pepper

8–12 sausages

1 tbsp vegetable oil

Method

Take a measuring jug or large mug and fill it to the half-pint mark or 300ml mark with flour. Sift into a large bowl and add salt and pepper to season.

Break six eggs or more until you reach the same volume in the jug. Add to the flour and mix. Then add 300ml of milk and slowly beat the mixture together until you have a batter with a consistency of double cream.

Leave the mixture to stand for half an hour.

While the mixture is standing, preheat the oven to 180°C/ fan 160°C/gas 4, then heat a large frying pan and fry the sausages on a medium heat until they caramelise and turn golden brown.

Pour the oil into an ovenproof dish and put in the oven for 5 minutes. Then add the sausages and pour in the batter. Immediately put the dish back in the oven and cook for 35 to 40 minutes until the batter is well risen and golden brown.

Serve with seasonal veg and gravy.

easy. We both felt sad – and that's precisely how it should be. Feeling that basic emotion is really important. It's the price we pay for understanding where our food comes from, same as other cultures, past and present, would kill an animal but then give a blessing or say thank you. There's another element to consider – we have taken great care of each and every animal we send to slaughter. We have given it the best life ever and now we want to give it the best death – the most painless and high-welfare we can find. Some would say better than that would be not to kill it at all, but that ignores the fact that the animal is part of the food chain – it's always been part of the food chain. It's not the cat sitting on your lap. Early on, Michaela seriously thought about becoming vegetarian. But the truth is that if people didn't farm rare-breed animals for food, then the gene pool would shrink to an unmanageable level and the breed would soon be no more. There is always that very delicate balance to consider.

Remember, like I always say, when you see meat in a farm shop, the question you should ask is not why it's expensive but why the mainstream alternative is so cheap. The only way a chicken can cost three quid in a supermarket, or a three-course Sunday lunch in a pub be a fiver, given the costs of rearing, feed, slaughter, packaging, distribution and sale, is by compromising standards. In the UK alone, almost a billion chickens a year are reared with such haste that many are so heavy at such an early age that they can barely walk. An RSPCA poultry welfare specialist told me once that, in human terms, a basic supermarket chicken would weigh 28 stone by the age of three months. The alternative might be more expensive, but then again you know exactly where the meat has come from, where that animal lived and what it was fed. That means quality of life for the animal, and sustainability, traceability and flavour for the customer. All in all, a low-carbon footprint alternative to the blandness of supermarket shopping. In 2020, the Farm Retail Association estimated that farm shop and café turnover was £1.5 bn. That's a lot of people backing belief with action, a reversing, to some extent at least, of the switch to pre-packed meat rather than buying from and using the knowledge of a local independent butcher.

Essentially, intensive farming, and the disconnect it produced, is another legacy of wartime Britain. The need for home-produced food meant a vast expansion of arable land at the expense of scrub, pasture and hedgerows. Post-war population booms and the development of pesticides, machinery and farming techniques meant the attitude of 'bigger is better', with its promise of never-ending supplies of low-cost food, never went away. It seemed like we'd been weaned on to chemical agriculture for life. Now we know the harm to the environment brought about by such methods and the repercussions for biodiversity, the mood has swung the other way. The future of the natural world is wholly tied up in the way we use the land and how we eat, which is why my television work extends to programmes about sustainability, lifestyle choices and food. That in itself leads to some very odd situations – I remember once filming in a baked bean factory while looking nervously at my phone as it filled with messages about a raccoon escape.

Spring, like all the seasons, is essentially just a matter of weeks, but undoubtedly the time when the most remarkable and life-affirming change happens. From frosty beginnings, by late spring everything is at peak fitness. I'll start picking out apple trees at the side of the road that I hadn't noticed before they started to blossom. At the same time, the hedgerows are painted white with their own blossom, blackthorn and hawthorn buzzing with bees and butterflies feeding on that bounty of nectar. I wallow in that heady scent, a sign that the hedgerows really have kicked in, a tangled mass perfect for nesting robins and blackbirds, all sorts of incredible wildlife. The RSPB estimates that hedges support up to 80 per cent of woodland birds, 50 per cent of mammals and 30 per cent of butterflies. That's not even counting the plethora of reptiles and amphibians, including our old friends the frogs and toads, which rely on this perfect habitat.

It's shocking to think then that more than 100,000 miles of UK hedgerows have disappeared since 1950, mainly due to changing farming methods which favoured larger field systems. Here, we've made a point of planting hedgerows and big thickets. When we built the exit road, for instance, we mounded the excavated earth alongside and planted oaks, between which hedgerows soon naturally started to form. That wasn't hard to do and yet it created the basis of an ecosystem that, like so many of our hedgerows, will, we hope, be around for hundreds of years. In doing so, we'll be joining thousands of other farmers who for years, long before we came along, have looked after our landscape, footpaths, hedges and fields. Considering the work they've put in, they aren't given half the credit they deserve.

While rare-breed and endangered animals are what drives us forward, we never forget our responsibilities to the indigenous wildlife. Hedgehogs are taken for granted. They certainly seemed to be abundant when I was a kid. However, whereas in the 1950s they numbered around 30 million, now there is estimated to be less than a million snuffling around our gardens and hedgerows. Again, intensive farming must carry some of the blame. Field margins have disappeared while insecticides have destroyed their food source. In urban areas, increased fencing has reduced their ability to wander and forage. Remember, you can always give our prickly pals a helping hand by leaving a plate of wet cat or dog food out and widening their territory by making sure there's the odd little hole in your fence.

We try to do our bit by working with a rescue centre to rehabilitate ill or injured hedgehogs back into the wild. We've basically built a halfway house in the woods, temporary accommodation, essentially a hedgehog Airbnb without the rent, while they reacquaint themselves with their natural environment. There's certainly no shortage of hedgehogs on the farm now. I don't care what animal it is, from the tiniest mouse to the rangiest deer, releasing anything back into its natural habitat will always be an incredible feeling. Our wildlife give us so much more than we can ever give back, but there will always be ways we can make a difference.

We do get a fair few animals left at the gate. Boxes of puppies or kittens aren't uncommon, same with injured birds, but we also receive a fair amount of young cockerels. Obviously, 50 per cent of chicks are male and people don't want them. They don't want to address the problem and so dump it on someone else. What tends to happen is you pick up the box, the bottom drops out, and there's cockerels everywhere. Leaving unwanted animals with us is clearly not something we encourage. It's not why we're here, it's not what we do.

If all goes well, late spring will also mark the arrival of reindeer calves, undoubtedly one of the most special moments of the year. Fortunately for us, the UK is just within the range of climates that reindeer can tolerate. Any further south and chances are it would be uncomfortably hot. As it is, we make sure they have a large tree to deliver shade in their enclosure while our woodland provides a welcome cooling-off area. Even so, they can't wait to get their thick double-layered coats off in the warmer months, the few raggedy days they spend shedding ruining their otherwise impeccable dress sense.

In the wild, reindeer are known to live in herds numbering more than 50,000 animals. Lacking an acreage of several hundred square miles, this is impossible to replicate, so our adults run to a rather more manageable three. Visitors are often taken aback to hear them click. The noise of a tendon in their foot passing over a bone, clicking is thought to be one way that a herd keeps together in poor visibility. It also sounds uncannily like rainfall – imagine several hundred reindeer clicking at once – and has led to speculation that this is where reindeer get their name. I'd love to think that's true but I suspect actually it's got more to do with the fact that in Old Norse they were called 'hreinndyri'.

It wouldn't matter how many reindeer we had, a birth, as with any of our animals, will always be an anxious moment. The joy of witnessing a calf wobble to its feet and stand for the first time is matched only by the relief. At that point, the baby's health is hard to check in detail because mothers are quite naturally protective and really don't appreciate humans getting too close. All we can do for a while is stand back and appreciate the little marvel that has arrived in our midst – if adult reindeer are one of the most beautiful creatures on the planet, then their young are even more so. We did have one calf, Ivy, which appeared to have a wonky hind leg. However, as a general rule of thumb, it's best to leave any animal mother and baby alone to bond and only intervene if there's an obvious serious problem. In Ivy's case, the leg righted itself and all was well.

Behind them in the woods, it's the wild deer fawns that are filling their lungs with that first gasp of crisp, clear air. While the wild deer exist separately from our wildlife park, occasionally the two worlds do collide. One recent spring, a lone and wandering roe deer fawn was found and brought to us. Female deer often leave their young while they go off and feed. Normally, that's not a problem because mum is always going to come back. Except this fawn's mum had been hit by a car. In the nick of time, the youngster was found by a passer-by, brought straight to us and we were able to take her in, feeding her by bottle and finding a space in with the goats, which at least were her kind of size, if not always quite so well behaved. I love a goat, but they're so naughty when they're little, especially good at issuing a quick and tear-inducing headbutt to the testicles. To our eyes, fawns have an incredible delicacy but, as with all wild animals, there is a robustness – there has to be. While rescue is part of what we do, we don't want to be home to any animal that could be living wild. However, we also need to recognise that a deer raised by bottle will lack the nervousness and fear it needs to survive outside our protective border – especially if it thinks it's a goat!

Injured animals always raise questions as to what to do with them. Find a swan with a broken wing and the answer seems obvious – have the wing repaired and off it goes again. But rarely is life so easy. What if the wing is repaired but not to an extent that the swan can still fly? What happens then? The answer can be that such a bird, mammal, or whatever, becomes a breeding animal or even simply a fantastic representation of its species, or nature as a whole. One way or another, wildlife parks, like our own, tend to find a solution.

When it comes to spring and the non-natives, the thing I really love is the butterfly house. We open its doors at Easter – the cost of heating it throughout winter would be horrific, with the add-on of burning fuel not being at all environmentally friendly. It's a calendar that works because a lot of our butter-flies are kept in the caterpillar stage over winter. Those that aren't head to other breeders for a holiday. Once the outside temperature increases and the days start getting longer, their home here, a little like a greenhouse, starts heating up natu-rally. It basically operates two weeks ahead of spring, so while the wild garlic is doing its stuff outside, inside it's all about canna lilies coming out, banana trees shooting up, and the bamboo becoming more vigorous. By late spring, I'm walking past a beautiful tropical pool accompanied by the fluttering of the most intricately carved and coloured wings. It feels so special, my own little wormhole, just like when I was that kid at Mole Hall – from the British springtime straight into the tropics. If there's anything that might tempt me to embark on another (probably mad) project it would be butterflies and their conservation. A biodome perhaps, home to an incredible array of these spectacular insects. I have these wild thoughts and then realise I can't keep up with my present everyday! There can't be room for another crazy idea – can there?

Many of our non-natives have breeding seasons throughout the year. In an unpredictable English climate it's up to us to time it so their babies are born in favourable conditions, pref-erably spring or summer. Any farm will try to avoid a winter birth because it's so much harder to deal with.

Over in our Outback Safari, a little slice of Down Under 10,500 miles away from the real thing, we have the wallabies, mammals which have evolved as a veritable breeding machine.

No matter how many years I've been here, there is one fact certain to stop any visitor in their tracks – the female wallaby has three vaginas. In fact, forget visitors to the wildlife park, the three-vaginas thing is guaranteed to bring anyone up short. I told Holly Willoughby on *This Morning* once and she was barely able to finish the show.

A wallaby's outside two vaginas lead to uteruses while the middle one is for giving birth. This evolutionary intelligence means they can store fertilised embryos for eight months, which in itself means they can pretty much permanently be pregnant. The gestation period is just a single month and at birth joeys make for a pitiful sight, only a centimetre in length and completely hairless. At that point of incredible vulnerability they set out on an unlikely journey, crawling all the way up to their mother's pouch where they latch on to a teat and settle in for the next seven months. Only when they have grown significantly larger do we get to see the ridiculously lovable sight of the joey poking its head, or, depending on its internal compass, foot, out of the pouch. When they do finally give up this warm, cosy and nourishing environment, they'll remain with mum for anything from three to nine months.

The good news is that, aside from the pandemic when numbers reached overflow point, with thirteen adults and umpteen joeys, we are able regularly to move wallabies on to other places where they can enjoy the space they need and potentially find a mate. It's at times like this, with animals being love-matched with others elsewhere, that I feel a little like Cilla Black. I've got three wallabies sitting on stools behind a screen and another looking for a date. Thing is, while the TV show's results tended to be a bit hit and miss, with animals it works a treat. When it comes to rare domestic animals, breeders operate as a network, engaging in programmes to ensure survival of species and expansion of bloodlines. Swap rare breeds for non-natives and it's one hundred per cent the same ethos, all about hooking up one animal with another and hoping it works. The difference between me and Cilla is she never sent her blind dates for a fortnight in a wildlife park. Maybe if she had there'd have been a modicum more success.

Moving a wallaby sounds simple, but in fact catching one is about as easy as knitting fog. Ours are Bennett's wallabies, masters of one of the greatest body swerves since Stanley Matthews. Those powerful back legs deliver an amazing hop, up to 13 feet, as well as quite a turn of speed – and a nasty kick to anything the animal doesn't like the look of. The whole highly impressive set-up is rounded off by a lengthy tail acting as a counterbalance. All this means is that realistically the only way to catch a wallaby is by becoming a human sheepdog. Persuade one or two other unfortunate souls to don the same disguise and then keep to a strict formation. Any gaps or lapses of concentration and your average wallaby will make you look a total mug. Once netted, they can be formally sexed, micro-chipped and sent off to pastures new.

Early on in our relationship with Australasian marsupials, we had a wallaby breakout. I put them in an enclosure where the fence wasn't the best, something spooked them, and off they went. Typically, we were abroad on holiday at the time.

'No problem,' I thought, 'I've got a general manager to sort this kind of thing out.' And then I remembered – she was flying out to join us. Handily, there's more than a few people we can rely on in a crisis and the fugitives were soon spotted a couple of fields away – in a country full of sheep, pigs, poultry and cows, escaped wallabies do tend to stick out a little. They might well have wished they'd never bothered with the vanishing act. By the time they got back, our emus, Alice and April, had decided their fellow Australasians' home looked rather pleasant and had invoked squatters' rights.

Nowadays, while not quite matching Glastonbury Festival's super-fence, we have a full-on perimeter barrier, with the wildlife park under strict supervision. Certain animals, however, are still capable of causing havoc within the park confines. Piglets, for instance, are always a bit of a nightmare. They have a knack for getting under wire and before you know it you've essentially got a gang of schoolkids running riot in the vegetable garden. For them it's all about opportunity. 'What can I find? What can I get into?' Thing is, with that smiley face, it's very hard to have a downer on them for too long.

To be fair, they're only copying the grown-ups. Our Saddlebacks and Tamworths especially have a habit of shunting large branches or logs onto the powerline for the electric fence to short circuit the supply. Given half a chance, they'll do it again and again. They also have a sixth sense as to when there's a power cut. As soon as those fences are down, off they go.

•

Goats, as touched on earlier, also have their moments, the issue being they have absolutely no respect for authority. Ask a goat to go through a gate and its immediate thought will be, 'I'm not doing that.' That might not be so bad were its second thought not, 'Much better to use the fence.' They put their front feet on it, bend the wire down, hop over, and that's that. We have a Bagot, thought to be Britain's oldest breed of goat, that gets out all the time. To make it extra annoying, it's always towards the end of the day. The public are gone, everyone's winding down, tired, and it's at that exact moment the Bagot thinks, 'Right, off I go!' Seconds later it's pulling the branches off the apple trees. Thing is, when goats get out, more often than not they very benevolently take their sheep pals with them. Again, I think the only reason they do this is because they can. They just don't care. No respect for authority whatsoever. More than once I've had to rather sheepishly (pun fully intended) collect both from another farmer's field.

Another time we had a bull run off. He was a very tricky character, only young, and became a bit freaked out during a tuberculosis (TB) testing session. That was it – he leapt the gate and legged it. I had to inform the police. 'There's three cows loose in the county right now,' they told me, which had me imagining whole gangs of feral cattle roaming around.

Where we are, with an A road on one side and a main railway line on the other, fence-hoppers really do have potential for disruption. I recall being a guest at the Royal Norfolk Show, slated to meet the president of the organising committee. For some reason, he was late. I was wondering what had happened when eventually he showed up. 'So sorry,' he said. 'The train was held up on the way to Norwich. There were some cows on the line just outside Ipswich.' My bloody cows!

Occasionally at the wildlife park we have albino wallabies, affected by a genetic condition which doesn't occur in the wild, resulting in a lack of melanin, the pigment responsible for skin, hair and eye colour. Their paleness means sunburn is a very real risk – there's a chance as you wander past the Outback Safari you'll be presented with the double-take sight of a wallaby having a coating of sun cream applied to its ears. Unsurprisingly, they're not massively keen. It's no coincidence that you don't see them on the beaches of the Gold Coast.

Albinos aren't as robust as their fellow wallabies, prone to skin and eye issues which put them at risk of infection. One of our saddest losses was Jenson, who had become a popular feature at the park. Jenson's breathing had become laboured and he was clearly struggling. We tried to save him by feeding him a high-nutrient diet, but his interest in food was no longer there. With no real sustenance, he went downhill quickly, despite the committed nursing of fellow wallaby Wendy, who had long groomed and looked after the old chap – a real education for us all in how a troupe will care for one another. We tried everything we could think of to save Jenson, including moving him to the tropical house to stay warm, but in the end it was all in vain, particularly upsetting for the rangers who had been in such close contact with him down the years. Wendy herself would sadly fall victim to a large abscess on her face – wallabies being prone to the condition – which effectively, awfully, left her with a hole in her face. Despite our every effort, she was unable to pull through.

Some people wonder if death is something I've had to harden myself to, but having watched and kept animals all through my childhood, feeding defrosted chicks to snakes, watching a dragonfly larvae grab a tadpole or a praying mantis eating a wasp, I soon understood that real-life nature doesn't quite match the images in the story books. That said, I have a great fondness for all our animals. The pigs are a good example. They were, after all, what started me off all those years ago, and to lose one is devastating. Older pigs can have heart attacks, the trigger for which can often be mating. It happened recently when we put one of our pedigree breeding sows, an Oxford Sandy and Black, in with a boar in the hope of producing one final litter. Oxford Sandy and Blacks are a tough breed but, like all pigs, lack sweat glands, which makes them susceptible to heat stress. Touchingly, the boar, Eddie, refused to leave her and had to be cajoled away before we could set about the emotional task of removing her body. There's no easy way to dress the process up. The sow would have to be incinerated – it's the required procedure and we have to follow it. The straw in her enclosure would also have to be burned off and the area sanitised. In practical terms, our job now was to nurse Eddie over the upset and back to breeding – life goes on, after all. However, while tradition states that you have to be tough to be a farmer, I'm not so sure that's true. Yes, you have to get used to losing animals, but that doesn't mean you don't feel it. I do, and I know plenty of others who are the same. Spend a lot of time with an animal and it's only natural you feel desperately sad when its time comes. Death, like life, is part of farming – it's an absolute and unwavering fact that just as we enjoy seeing life entering this world, then so we must also experience the flipside of the coin – but I'm sure, however long I do it, I will always feel the same.

I feel also the pain of our staff acutely. Nothing better illustrates their incredible levels of care and consideration than the hurt they go through when the time comes to say farewell. People often comment that it comes across on the *Jimmy's Farm* TV series just how emotionally involved our staff can become with the animals. It's as inevitable as it is true. Working here will always be more than a job. Looking after an animal every day, nursing it through illness, watching it give birth, or its family grow, leaves an indelible mark. That attachment is often hugely positive, but there are also times when it leaves you on the floor. For me, you have to be honest and show the rough and the smooth, otherwise you paint a false picture of dawn-to-dusk fluffiness. Because I have travelled to see wildlife, I know that some nature documentaries don't portray reality, because it makes better TV to ignore the truth. I have watched, for instance, numerous programmes that display the Serengeti as a vast wilderness that goes on undisturbed for mile after mile; just raw landscape and its nature. But then I went for myself and found that actually it's criss-crossed with fences. People live there in villages and towns. It isn't this huge empty space after all.

Edit out key elements of a story and it soon becomes something it's not. In our case, that means showing nature red in tooth and claw. When you have livestock, you also have deadstock. Animals do die and so we all take our seats on that emotional rollercoaster. We have to show those ups and downs.

Many people will, I'm sure, understand when I say it's the death of a dog that has affected me most. My little border terrier, Cora, had been by my side for seventeen years, right through starting out at the farm, running alongside the Land Rover as we set out to feed the pigs. So attached to her was I that I actually named one of my daughters after her, one of those things that seems a good idea at the time but can very easily lead to the kind of confusion more generally seen in a West End farce. When my faithful four-legged pal passed away, for instance, they could tell in the office I was feeling down about something.

'What's happened, Jimmy?' someone asked.

When I said that Cora had gone, I should have realised from the looks on their faces that they'd got the wrong end of the stick. I was trying to put a brave face on it, talking quite matter-of-factly about how I'd keep her blanket for old time's sake, perhaps a couple of toys, and they were staring at me like I was mad! Thinking about it reminds me of another time when I announced to everyone that the Prince of Wales had been in touch and wanted two boxes of sausages. It caused no small amount of amazement – *Prince Charles wants our sausages!* – until I explained it was the Prince of Wales pub.

Dogs have a big effect on your life. For a long time after she'd died, I used to think I'd seen Cora, not because I was seeing ghosts but because she'd made such a big imprint on my life. My brain was playing with that familiarity. In fact, it wasn't just my life she made an imprint on. Cora had a habit of sitting on my lap in the car. I was being driven through slow traffic to a photo-shoot in London, endless stopping and starting, Cora bumping back and forward on my knee as she looked out the windscreen, when I looked down and spotted the most perfect brown dot on my favourite shirt. Every time her backside bumped against me she'd been adding her own little design to the pattern. I don't think it ever caught on as a fashion trend.

With Cora gone, now we have Whiskey, the Irish terrier. I thought Whiskey was a great name – until she wandered off and I was left roaming the lanes shouting 'Whiskey! Whiskey!' in the dead of night.

In the big picture, I find it strange how death is something from which we try to shield ourselves. At some point, we all have to face the departure of loved ones. And yet even when it's someone who has meant so much to us, who has brought so much joy into our lives, has been at the heart of our very existence, we go into reserved mode. We have a funeral, a reception afterwards, and everyone sits around chatting quietly with a cup of tea and a sandwich. It's a set-up that by its very nature makes us feel awkward. Compare that to other cultures where they make grief about openness and celebration.

When my dad died, just two years into retirement, from mesothelioma, a cancer related to asbestos, clearly a result of his working on building sites, I found it, as I'm sure many will recognise, extremely hard to deal with. His death came just three weeks after the birth of Bo-Lila and at a time when I couldn't help but feel that a man like Dad, who had worked so hard all his life, should have been settling down with Mum to enjoy some long, relaxed and happy days together. My brother and I both read eulogies at the service. Anyone who has done the same will know that to do so is as important as it is difficult. I remember thinking, 'How do I get in his mindset?' But I recalled that he was a big fan of the great Irish comedian Dave Allen. He also liked to have a beer in the evening. So that's how I pictured him in my speech – sitting in the sun, having a beer with Dave Allen. I also described grief as the price we pay for love. At some point, you have to pay the boatman. The fact that we were all paying such a high price showed just how loved Dad was. Our pain reflected just how much we loved him; how happy he made us.

In the wildlife park, the cyclical nature of our work always tends to bring us round after a farewell. Following the loss of Jenson, soon we were welcoming a new wallaby male, at which point attention could turn once again to renewal.

In his Outback Safari home, the newcomer would join the emus. The world's second largest bird after the ostrich, emus, with their long necks and legs, can grow as tall as 1.9 metres, travel great distances, and reach speeds of more than 30 mph.

Ours had no chance of living that kind of life – they were being kept in someone's back garden shed. The male emu is unusual in the avian world – whereas those of other species enjoy nothing more than displaying their machismo through the medium of colourful or intricately patterned plumage, it prefers to hide its light under a bushel. To find out if an emu is male, we rely on blood analysis. Well, usually. During the pandemic, when lab services were restricted, we had to resort to a more basic approach – an inspection of the cloaca, the chamber into which the intestinal, urinary and genital tracts open. It's present in all birds as well as snakes. I'm far from inexperienced, having worked with and around animals of all shapes and sizes all my life, but even so, the realisation that I needed to familiarise myself with what an emu's phallus or clitoris looks and feels like came as something of a shock. As did the thought of wrestling with five examples of prime emu, each weighing 100 lbs, with razor sharp claws on each foot. How the emu didn't get a walk-on part in *Jurassic Park* I'll never know. After a bit of a wrestle, and a pair of jeans now sporting that oh-so-trendy ripped look, we'd established we had four females and, positively, for breeding purposes, a male. That was it, job done, for ever – our old friend the microchip means once sexed never repeated. The rubber gloves can stay firmly in the box. Looking back, it's one of the strangest experiences I've had at the wildlife park. I can well understand why emu sexing isn't something that often features in the pictures on the calendars found in gift shops.

•

Meerkats are a little easier to sex – turn them upside down and have a look. They are also prolific breeders, with females giving birth to up to five young after a pregnancy of around eleven weeks. We started out with two slender-tailed females, Kalahari and Timone, domestic pets which we rehomed. They bred and soon we had fourteen.

The meerkats were the first very obviously non-native animals we had. Exhibiting them meant a massive jump in mindset as well as commitment – not only would we need to take on new staff but they would need different kinds of expertise. We were also opening ourselves up to a whole new raft of rules and regulations.

Ignore the East European accents of the popular animals of TV advert fame, meerkats are actually found in the African desert. A type of mongoose, they live in large family groups known as mobs, highly appropriate considering their mafia-like insistence on working as one to ensure the survival of all members of the clan. However, it's more of a Godmother than Godfather set-up, with the dominant female issuing the orders. Some meerkats are put on watch duty – they bark if they spot a predator such as an eagle or snake – while others must hunt for food, and a few are given the less glamorous task of using their large foreclaws for burrowing duty. All, however, contribute to the schooling of the young in traditional meerkat ways. Because Kalahari and Timone had been domesticated, they lacked those learned meerkat skills. They weren't inquisitive about their outdoor space and had no interest in digging or sentry duty. Fortunately, they both listened to Mozart – not the composer, rather the male who came in to breed with them. Mozart clearly saw his duties as going beyond the purely physical and revealed educational skills that would be the envy of anyone who home-schooled during the pandemic period – vitally important because Kalahari and Timone could then pass on those abilities to their young. The chain of behaviour from one generation to another, broken by someone's desire to domesticate a wild animal, had been re-established.

One meerkat, however, would come to exhibit behaviour totally unlike any of his companions. His endless desire to break out of the enclosure soon gained him the nickname Steve McQueen, in recognition of the Hollywood actor's role as Captain Virgil Hilts in *The Great Escape*. While the German guards would contain Hilts in the punishment block, we don't, of course, have such a structure. Time and again, no matter what we did to make his home escape-proof, he'd somehow manage to climb the wall and be off, while several of us tried to track him down and then capture him in a net. I would hate to think how many staff hours we have spent pursuing Steve. It really was very difficult to think anything other than he was having a very big laugh at our expense. Even when we trained cameras on him to work out his methods of escape, and modified the walls accordingly, he took it in his stride. 'Is that a rounded anti-climb edge you've just fitted? No problem. See ya!'

We will get the better of Steve yet, however. We have just built a bigger enclosure which will offer a better environment for both meerkats and visitors. We are also confident it is escape-proof. I'm sure Steve is already planning ways to prove us wrong. We're making sure he has no access to a motorbike.

The meerkats are one of the animals that children always make a beeline for. It's obvious why – with their stripy tails and distinctive dark patches around their eyes, they make for an irresistible combination of comic and cute. Truth is, though, they can be fierce characters when they want to be. Many's the scorpion that's rued the day it took a meerkat on. They are also known to attack and repel snakes that mistakenly see them as easy quarry. The biggest threat to meerkats, as with so many animals, is climate change. As temperatures rise in their native Kalahari Desert, so do their chances of suffering serious dehydration, affecting muscle and growth. Desert-dwellers have long adapted to survive heat, but there's only so hot that any animal can go.

Of course, you don't need to be in the Kalahari for global warming to have serious ramifications. Spring here is unarguably coming earlier, which presents its own problems. If suddenly you have a mass flowering, and the bees aren't yet out, missing that nectar source can have a catastrophic effect on their survival. Meanwhile, because our animals are so tied into the seasons, shifts in weather patterns can throw breeding habits out of sync, while, as with the bees, much-relied-upon food sources are no longer there.

In the time I've been here, the decline in insect numbers – the sudden shift in weather patterns playing havoc with their life cycles – has been one of the most noticeable indicators that change is afoot, a sign of something going very badly wrong in a very short space of time. After all, insects were around before the dinosaurs, before even grass and trees. There's a lesson there – listen to the little people. They're by far the best at warning that something is going wrong.

Hopefully, the spring we still largely recognise will continue to be an incredible companion, leading us on an astonishing journey. Look one way and you see the struggle for new life, look another and birth and rebirth are there in all their many magnificent forms.

Along its winding road, spring hands us gift after gift, but eventually it must reach its end. The good news is that the beautiful dream it has been selling along the way is about to come true.

Summer

Big skies, even bigger sunsets and, fingers crossed,
a wildlife park full of contented and well-behaved
(yes, Barbary macaques, I am looking at you!) animals.

An urban myth persists that it's unwise to visit a wildlife park in summer. The animals will, it is said, be lying around in the sun doing nothing. Tell that to anyone who's seen our tapirs rolling around in their pond, submerging themselves for minutes at a time with only their trunks popping up as snorkels. Or our kookaburras, dust-bathing or down by the pond eagerly watching for lizards or dragonflies – gone in a flash if they're unfortunate enough to find their way into their field of vision. OK, the meerkats might have a little sunbathe, but anyone who has seen three or four meerkats lounging on their backs, lacking only a piña colada and a Jackie Collins novel, will know it's one of the best sights you can ever see.

Just like for us, summer's a time for our animals to really love life and live it. However, in 2020 when Covid hit and lockdown was implemented, I wasn't alone in noticing something very strange happening on the wildlife park. The behaviour of some of the animals notably changed. It felt like the fewer people they saw, the more withdrawn they became, as if somehow they shared the feeling that the world had shut down – as if they experienced the same anxiety.

Jerry the alpaca was a case in point. Jerry likes to make himself known, more than happy to wander over and give a stern stare to anyone he feels needs bringing down a peg or two. He's very confident in his own skin, the Cristiano Ronaldo of the alpaca world (although I always think he looks more like Ken Dodd), and it's one of the ways he keeps himself amused. Everyone needs a hobby and that's his. But with no one to intimidate – not only did we have no visitors but with staff furloughed we had only eight essential workers present – Jerry reminded me of someone addicted to social media who'd suddenly lost their phone. He'd come running over to me every time I appeared. Disappointingly for him, I was lacking in idle gossip.

We also had a sheep which had enjoyed being fed by the public rather too much and was now disappointed to find itself on a sudden weight-loss regime. It was obvious too that tapirs Tip Tap and Teddy were missing their adoring public. Desperate for attention, they suddenly became extra partial to a tummy tickle. That was another surreal moment – 'Hang on, the world's in the grip of a pandemic – and I'm here tickling a tapir's belly!'

Other animals, the shy and nervous ones, such as the skunks and raccoons, took the opportunity of empty paths and walkways to come out more than ever before. Skunks are naturally nocturnal. Like Dracula, at sunrise they retire to their place of repose. Unlike Dracula, they express displeasure through the use of anal musk glands, about the size of a grape and surrounded by powerful sphincter muscles. When suitably incited, a skunk is capable of emitting a concentrated stream of spray as far as 4.8 metres. I have yet to get in its way. I really do doubt the washing powder has been made that's capable of addressing that particular odour. It's not a claim I've seen made on any adverts.

Our two skunks, also rescues, having been kept as pets, got out recently and had a lovely little jaunt around the wildlife park before they were caught. One has the ability to spray, the other doesn't – 'pet' owners have the glands removed. Ranger Stevie caught the one that doesn't spray. Ranger Sophie drew the short straw and went after the other. Unfortunately for her, the skunk wasn't desperately keen on being recaptured and went full-on with the scent, some of which went in her mouth. Do please feel free to put this book down to go and be sick now.

Because of their malodorous reputation, skunks hold a fascination for children, helped by the fact that they are very cute at the same time, as are so many other members of their family – otters, badgers and the like. Ferrets too – when they're babies especially there's not many things cuter than a ferret. Scorpions might not be cute but they fascinate for the same reason – that sting in the tail, that ability to react when pushed.

It's worth mentioning that a skunk will only spray as a last resort. It's a defence mechanism and they will display their tail beforehand to say 'Back off!' The problem for Sophie was she didn't have a choice in apprehending the escapee. I think I can confidently say that few Ipswich residents would welcome a skunk setting up in their shed.

I expect the skunks and raccoons would have loved the wildlife park to stay quiet for ever – no public, no wedding parties. Then again, they aren't the ones paying the bills!

Others, the coatis, meerkats and goats included, saw lockdown as a chance to be heard. A lot. Endlessly noisy, they seemed to come at the situation from the standpoint that if there's a silence, fill it.

The native wildlife, meanwhile, clearly saw the lack of visitors as a chance to have an explore. A fragment of goose egg provided a clue that a fox had seen a chance to gather a rich bounty. Never underestimate how quickly wildlife will adapt to change. Foxes are a great example of that ability. Give them a chance and straight away they'll take it.

Elsewhere, stoats suddenly became very visible, like word had got about that they could wander around with impunity. On several occasions, I'd see them running up and down the paths, and again, like a buzzard soaring on a thermal, it was hard not to think that there was anything more to their behaviour than simple enjoyment – 'I can do this, so I will!' By the way, if you're ever baffled as to whether you're looking at a weasel or a stoat, look for a black tip to the tail. If it's present, you're looking at a stoat. They are also bigger than weasels.

What we do here is a business and so suddenly to be denied our main source of income – the public – was alarming.

Margins are tight at the best of times without losing revenue from the farm and wildlife park, shop and restaurant. In food and farming there are always significant moments of trial. Any animal-health issue where there's danger of being quarantined or closed down is always a worry. The foot-and-mouth crisis of 2007 and banking crisis of 2008 with accompanying recession were two all-consuming whirlpools we'd had to battle to keep our heads above water. There had also been national outbreaks of bluetongue and avian flu. If it's not disease, it's the weather. In 2013, East Anglia suffered terrible flood and storm damage. At one point, we were without electricity for four days – not the best of news when you have a shop and restaurant, and sure enough our stored meat was ruined. Not only could we not freeze anything but, this being autumn, we couldn't keep people warm either! Then, in February 2022, we were visited by the violent Storm Eunice. The Suffolk Punches' stable was blown into another farmer's field, the entire back fence of the monkey enclosure was demolished, the coatis' fence disappeared, a pig hut flew through the air and the barbecue area's roof took off. Oh yes, and a big tree came down and smashed through a fence in the reindeer enclosure. Luckily, all the animals had been safely locked away because of the high winds. However, there was one that became airborne. 'Pigs might fly' they say – well, ours actually did! We have a ginormous black-and-white plastic piggy bank we use to raise money for the Rare Breeds Survival Trust. It took off and landed next to a digger belonging to a builder doing some work for us. 'Right,' he said, 'time to go home now!' He had a point. When a ten-foot piggy bank lands next to you, it's definitely time to pack up for the day.

My attitude with the farm and wildlife park has never changed – whatever the problem, there's always a way out. There's always something you can do. When, early on, we were in financial dire straits, we created pop-up markets in village halls and, as mentioned earlier, started our own market at the farm. At the same time, we had the shop open seven days a week and were attending every farmers' market we could find. Had we not done so we would have faced the very real threat of insolvency.

My belief has always been that nobody is going to make you or your venture successful – you have to get out there and make it happen. I'd try anything I could to drum up business. When we went to the Food Fair at the NEC, I got Asa to pretend he was a customer. We had a very loud conversation in which he was *very* excited about our products and I professed, with equal volume, that if he wasn't quick they might very well run out. It did the trick and soon there were plenty of people coming over to see what the fuss was all about. If my memory serves me right, a bit later, when things had quietened down again, Asa put a different hat on and we did the same thing all over again. There's always a way! And there's had to be – even two or three years ago, we were feeling that there was a constant struggle just from the normal everyday strain of running a business. But Covid was on a different level. At our busiest time of the year, we were totally empty. Like any seasonal venture, spring and summer is the harvest that keeps us going through the more challenging months. Even a week-long shutdown at peak time would be a serious blow, let alone one going on month after month with no specified end date. The frustration was heightened even further when retail outlets were allowed to reopen but open-air attractions such as wildlife parks were left with locked gates. It's not just us. Seventy people work here – an entire network of great individuals, and their families, reliant on this small segment of East Anglia to keep their world spinning. Think about that too long and it can become overwhelming. Then there was the frustration of those desperate to get out of their homes, to come to a place like this where they could escape into nature, one of the best antidotes to the suffocation of lockdown.

Like most people, I'm not immune to anxiety, although perversely, rather than a global pandemic, it can be matters of the smallest concern that really nag at me. There are times when, if I've got an event in the middle of the week, I can't rest in the days beforehand. Same if I've got an evening function, my mind tunes itself to thinking I can't relax until it's finished.

Sometimes I have anxiety dreams. There's a recurring one where I'm taken back to studying for my PhD, in particular a piece of coursework I didn't bother doing. Right at the last moment, my academic laziness is found out – 'Doherty! I expect you thought you'd got away with that one!' Slightly more disturbing is the dream where I'm trapped in a big old mansion. There's one room that hates me, and of course I just happen to be in there when the door slams. Please, don't have nightmares!

Other times, a problem will jolt me awake in the dead of night, like the need to screw a branch to a wall, otherwise a monkey might fall off. In fact, I have a pad by my bed so I can write ideas down and take them off my mind. In the big scheme of things, these are trifling matters, and yet they can sit more weightily on my shoulders than the real pressure – the fact I am responsible for several hundred animals and an awful lot of people. It is, after all, my name above the door. In the end, experience, a reservoir of knowledge that somehow we always seem to survive, is what I draw upon most of the time.

That's not to say there aren't times when I really do feel like I'm in one of those old jungle explorer films, about to feel the sinking sands closing over my head. I recall one bout of absolute exhaustion where I had flu, was filming a TV series, the business was struggling and, to put the tin hat on it, the weather seemed to be week after week of relentless misery. I was in the office for ten minutes and did the classic slide down the wall, crouched there, head in hands – 'I'm so tired! It can't get any worse!' And then, in true sitcom style, someone walked in and told me something that was indeed much, much worse – a roof had fallen in – and the sheer ridiculousness of it all just made me laugh my head off. (Note: Before I had an office, in times of extreme stress I used to scream into a cupboard!) If you don't want to rely on a roof falling in to deliver that release of pressure, there are simpler ways to do so. Not reading work emails when you're supposed to be relaxing is one. Writing potential issues down is another. In black and white, rather than careering round your head, things can look so much clearer.

Possibly my Essex upbringing, coming from a family never slow to take the mickey, helped me out on the humour-as-a-pressure-release front. While my dad's side of the family was quite serious, my mum's side were always laughing. My mum and her mum together were terrible. Anything could set them off. Someone in a big pair of glasses at the door would be enough to have them rolling around. It's a bit of a feature of the county to be able to laugh at yourself and whatever calamity you find yourself in – I've done a lot of laughing down the years! And I'm glad. It's a great way of calming a situation, putting people at their ease. As long as you can laugh, you have a chance. It delivers an immediate dose of perspective, not just for you but for those around you. I'll even do it if I'm on my own. I was cooking recently and spilt a load of fat down myself. Don't get me wrong, there was a fair bit of swearing, but there was a lot of laughing as well. The only problem now is that my kids like a laugh too – more often than not at my expense. I see them out the corner of my eye formulating jokes or pulling funny faces. It's infuriating. You're battling the monster you created! They've got all your powers.

Right from the start, there have been times when we've had our backs against the wall. Forget the animals, if the knack-erman was coming for anyone it was us! In our first winter, stymied by the shop opening later than we had planned, we had a deficit of around £30,000. Michaela and I weren't exactly hardened business types at that point – I was an insect expert and she worked in television. I expect we'd have got short shrift on *Dragons' Den* – 'Rare-breed pigs? I'm out!'

Thing is, if the toilets have flooded, a fence has gone down, or a pig has taken up residence on the A14, you can either look at the whole mess and start screaming and shouting, 'Disaster!', possibly while running for the hills, of which there aren't many round here, or you can pick up the mop, get the hammer out, and head for the main road with an armful of apples. Do that and all of a sudden it's problem over, done. Even the apparently most insurmountable issue can be overcome. Look at a conundrum as a cliff-face and you've got no chance. Look at it as a series of steps to the top of the mountain and you've got a chance. Like evolution itself, making progress is far from a single massive leap, it's a steady, and sometimes long, climb. You must also at all times remember why you are here. In my case, it was to farm and produce food in a rewarding and sustainable manner, which in itself would produce a complete sea change in lifestyle. Money doesn't have to be your goal. If you're not chasing your next million, then it's easier to be at peace with yourself.

There's something, no matter what the external pressures, we always have to remember. What the business is really reliant on is the animals. They are what makes us tick. By far the biggest cog in the mechanism. Looking after their physical well-being – food, health, comfort – is the absolute bedrock of what we do. Covid, and this odd situation of their very public environment being reshaped into one without people, reinforced the need to keep an eye on their mental well-being too. Any animal that experiences change is subject to stress and anxiety. Early on in our tenure, for example, we brought in a boar called Blaze. The move to pastures new seemed to have unsettled him a little and he wasn't even remotely interested in the sows – very unusual boar behaviour indeed. In pig terms, he was lonely and depressed. To cheer him up and give him a bit of 'company', we put a radio in with him. With a bit of trial and error, we came to see that while Classic FM wasn't his thing – he displayed his distaste by attempting to smash the stereo up – a bit of Radio 4 very much was. Over time, he became a very big fan of *The Archers*. *Money Box Live* not so much.

It's up to us to make sure we're all over any welfare concerns and do what we can to help. That's the beauty of having rangers who know the animals and their behaviours so intimately. We've always been very lucky to have highly committed and knowledgeable staff across the board. We work together to make a very complicated set-up work.

When we need specialist know-how, we recruit from within the wildlife park industry, but as time has gone on, we've also begun to train in-house. I'm a patron of Suffolk New College, in Ipswich, and also involved with Writtle Agricultural College, in Essex, so there's no shortage of enthusiastic young people studying animal science and care who we can take on as apprentices. If we can't then offer them a role, we can often find them a job elsewhere, but it's always lovely to have staff grow with the business. We often get really positive emails about our rangers – 'they explained so much about the animals', 'they were really happy to show us round'. It always means so much to see their efforts appreciated.

When Prince Charles came down, it was a real pat on the back for every one of our staff, recognition for the incredible work they do, and we tried to make sure he met as many of them as possible.

That day was also incredibly funny. For a start, my daughter Cora seemed to be under the impression it was an entirely different person who was visiting – 'Is Prince Charming here yet?' Then, with the visit being unannounced, all I could hear as we walked through the restaurant was, 'Look at that man – he's the spitting image of Prince Charles!' He was more than happy to say hello. I can imagine it was a bit of a shock – one minute you're having a cup of tea and a flapjack and the next you're chatting to the heir to the throne. Then there was the little girl who broke all royal protocol by giving him a massive hug and asking him if he wanted to stroke a chicken – the picture being one of the few times we've ever appeared in *Hello!* My mum did her own bit of protocol-breaking. She managed to say hello twice – sneaking into a second meet-and-greet line-up. As I watched him leave by helicopter, it was one of those moments when I couldn't help but think back to the days of showering in builders' bags.

Our team is always evolving. In the early days, having been here from day one, I found taking people on, delegating key roles difficult. It's like handing over something precious to a stranger – your big worry is they won't take care of it. They haven't got the same degree of attachment – it's not theirs. The solution is integration. Allow people to dive into a business and it soon becomes something they care deeply about, a job they take pride in. You can also still stick your head round the door and remain involved in that part of the business – much to the annoyance of the person whose role it is! Sometimes delegating works, sometimes it doesn't. Put the wrong person in a position and it can be a disaster, but we've always been astute – or lucky!

The business side of the park is the hardest element for me. It's easy to get excited about an emerald swallowtail butterfly, originally from South-East Asia, emerging from its chrysalis, taking its first flight and drinking nectar from a flower. But how much excitement is there to be gained from the depreciation of our vehicles this year? Or the latest VAT regulations? Or the price of a new septic tank? Doing up the loos doesn't have quite the same appeal as a shipment of macaques, but it still has to be done. I have endless meetings with accountants where I just end up watching robins out of the window. Other times, I'll see starlings forming a murmuration – good luck trying to get me to concentrate then. But without those meetings, I can't watch the emerald swallowtail. The two are inextricably linked.

I'm lucky that Michaela takes so much of that pressure away from me. She is instrumental in running the business, overseeing the staff, dealing with the nuts and bolts, as well as being a massive creative force, especially around the gardens and the planting. It's funny how the commonly accepted narrative has come to be that I'm the more erratic and instinctive half of the partnership who makes things happen, while she's the sensible one. Michaela has actually been the catalyst for so much change. She's brilliant at seeing the bigger picture, of how an idea can actually be expanded to become so much more exciting, long-lasting and worthwhile. That perspective has always been invaluable. When you first put your flag in the soil, it's all too easy to be conservative and cautious. Michaela's attitude would be, 'Hang on – why don't we do this instead? Why don't we go for it?'

Covid was a pivotal time. There's lessons to be learned out of it – both good and bad. It's one of those periods that is going to be part of history, that changed the world, that stopped the world, and it's foolish to think we can just brush the whole thing off and carry on the same. As a family, we tried to find the positives where we could, and like so many others, we discovered what rescued us more than anything else was a reconnection with the natural world. We were as desperate as anyone to get outdoors. That might sound strange for a clan lucky enough to have such a wealth and variety of wildlife on its doorstep, but a lot of the time the demands of normal life can get in the way. There were very few silver linings to Covid, but, as time went on, that ability as a family to lose ourselves in the farm and wildlife park became very special, to the extent I find it hard to see the summer sun without thinking about it now.

With the gates shut, the only children on site were our own. Molly Rose, Cora Mae (her full name), Neve and Bo-Lila all love animals. Naturally, as I certainly would have done, they thought having their own wildlife park was wonderful and took every chance, clad in their mini-uniforms, to help with the feeding, cleaning out and all sorts of other jobs. In fact, considering we were down to a skeleton staff, their free labour was much appreciated!

Around nature, kids especially see the world through a different lens, something of which I became acutely aware as I watched the girls become more and more immersed in the animals' lives. I noticed how, just like my own childhood self, the more they saw the more they loved the detail. The wallabies, they'd remind me, aren't just wallabies, they're Bennett's wallabies, known for a distinctive red tint on the back of their necks. It didn't stop with the wildlife park. We were watching a wildlife programme when up popped what looked like a manatee. Immediately, one of the girls pointed out that it was actually a dugong (dugongs never leave saltwater; manatees live in salt and freshwater – remember that next time you're stuck for conversation in a lift). Molly Rose developed a habit of flicking through David Attenborough's *Life on Earth* book, showing me a picture and asking me, 'What animal is that?' She loves accruing knowledge – and testing mine. At times during Covid it felt like they were home-schooling me! At that age, children are like sponges, soaking up information, and it's great to see them revelling in newfound information and wanting to pass it on.

On a practical basis, the closedown meant a chance to address so many of those jobs that sit on the to-do list and somehow never seem to be completed. The snag was that the usual materials for building and repairs were pretty much unavailable. But then I found myself really quite enjoying the idea of seeing what we had lying around that we could use instead. It reminded me of being a kid. Nipping down to B&Q for a few hundred quids' worth of wood and wire wasn't an option for my teenaged self and so I was used to making do with whatever I could find to construct various enclosures (with varying degrees of success).

One of the most incredible developments was the guinea pig village, surely one of the best examples of rural housing in the country (Prince Charles's Poundbury village has nothing on it), complete with five dwellings, church, castle, hospital and pub, The Groundhog Inn – unless you want a bit of straw, a dish of water and a carrot, I wouldn't recommend dropping in. Thirteen guinea pigs moved in, and when the park reopened, the village soon became a popular attraction. I'd advise timing your visit well. At weekends, for example, you might not see the residents if they're at Sunday service.

Something quite poignant happened during work on the guinea pig village – at one point, I looked down and realised I was using my dad's club hammer. It was two years to the day since he died. Reminders of Dad come from anywhere and everywhere. Back at the start, when we were getting up and running, he would help out with jobs around the farm. There are still odd little things here that he worked on, possibly with the very tools I still use now, such as the tiles on the roof of the shop. The circle carries on. That's the thing with having a dad who was laying bricks right from the time he was in short trousers. Inevitably, there are buildings, walls, houses, everywhere that he's responsible for, each, in its own way, a little memorial, one that will be there, hopefully, for many years to come. I actually have some of my great-grandfather's tools too. It's quite something to think of them in his palm, and then in my dad's palm, to the extent that rather than holding a hammer or a spanner, I'm actually holding their hands as we jump across the generations. It's strange how certain elements of people live on after death. I often find myself saying things my dad might have said. Or I'll catch myself in the mirror and recognise a facial expression.

By the time Dad passed away, the wildlife park was well developed. He never really let on much in terms of being proud of me and what I'd achieved. Mum's a little more effusive – 'Oh, it's marvellous!' she'll say. But they were always more likely to tell someone else of their happiness that it had all come good rather than me – that old working-class ethos that no one should get too big for their boots. That's their upbringing mingling into my upbringing, but I like to think that the underlying feeling was that my younger years of requisitioning every spare inch of their home to house my animal collection did actually pay off in the long run, even if it did mean the occasional escaped python hanging from a curtain rail.

Billy Bragg wrote a deeply touching song about the death of his dad, 'Tank Park Salute', referring to the one-armed tank turret impression they shared when he was a lad. Listening to it inevitably transports me to that little boy being carried up to bed, my dad reading *Treasure Island* to me, those big adventure books. I'm pretty sure that Dad, despite the occasional word to the contrary, the odd complaint about the fish tank, quite liked that I had all those animals as a kid. 'He's animal mad, that kid!' he'd say. Underneath, I think he was pleased that I was genuinely so fascinated by something. Only when you're a parent yourself do you really see what parents see and maybe, no matter what I took on, no matter what I did, he could always see that kid in me. He'd be right to have done so, because that kid is absolutely still there. He just hasn't the time to indulge himself like he used to, although he's still very easily sidetracked, at the moment by a jumping spider that lives on the landing. Jumping spiders get their name from their natural agility, useful in hunting prey. You can really see their brains ticking over, calculating. They will actually turn and look at you, which I really love. If they then see a prey item, they'll get a little bit closer, work out the distance, and jump on it. As with any house, spiders are everywhere, but I've watched this one grow over the year. More often than not, if I'm using the upstairs washroom, I'll check if he's still around.

It's funny how those quiet days of the pandemic mentally cast me back down the years, and not just to my childhood. With just a few deeply committed people beavering away, I was reminded of the challenges we faced in our first summer – all hands to the pumps as we battled to complete the list of daily, weekly and monthly tasks that seemed after a while to blur into one. Maybe that was the point I should have got camels in. Not only would they have found the hot weather a breeze, but they would have come in handy with all the dragging and lifting.

I'm not sure our present-day Bactrians would be massively pleased to be put to work. Even at the best of times, Alice and Arthur tend to look a bit miserable, very straight-faced, like sentries at a royal palace. If they moved in next door, you wouldn't be expecting an invite to a wild house-warming party. They're not a laugh a minute, that's for sure. Thing is, I get the feeling they're not quite as fed up as they look. They just don't want to be seen to be happy, like if they heard someone break wind, they'd turn away so they wouldn't be seen smiling. Their downbeat aura is let down a little by their vast Lily Savage eyelashes, there for a reason – living in the desert isn't about skipping around like an excitable schoolchild, it's about dealing with sandstorms. Add that to the constant need to preserve food and water and it's hardly surprising they're rarely mistaken for Timmy Mallet.

The arrival of Alice and Arthur signified another major point where it really felt like we were changing from being a rare-breed farm to a wildlife park. Whereas I'd been used to pulling up in the morning and seeing a bull, a cow or a sheep, suddenly there were these two great big camels. It took a good while to get used to that. For a long time, it really was 'Wow!'

There's an easy way to remember how to tell a Bactrian camel from a Dromedary. A letter 'B' has two humps, 'D' has just one. From the deserts of Mongolia and northern China, wild Bactrians are a critically endangered species with a population thought to be less than a thousand. While domesticated numbers of this two-humped species are relatively high, it's important that we do what we can to highlight the threat to those trying to survive in their natural environment.

As with all our animals, we carry out regular health checks on our Bactrians. Initially, we were concerned with Arthur's size. He seemed a little on the small side (for a camel). Take your cat or dog for a once-over and chances are the vet will stick it on the scales. If you want to see how your camel is faring, the same rule applies. Now I don't know about you, but my bathroom scales weren't designed with camels in mind. Even if they were, it would be a devil of a job getting Arthur up the stairs. Thank goodness then for the invention of the portable weighbridge. Once Arthur had been persuaded to put all four feet on it, he clocked in at a satisfactory weight and we could all stop worrying. Well, about that anyway. I'm still a little confused about how he's going to mate with a female twice his size. I'm sure that, as they say, love will find a way. Alice will sort it out. She's obviously the boss.

Annie and Annabel, our Suffolk Punches, the oldest English working-horse breed, dating back to the 1500s, make Alice and Arthur look like mere lightweights. Each of these beautiful chestnuts weighs the small matter of up to a ton. Their heft and power reflect the Suffolk Punch's long tenure as the backbone of farming in the east of England. For centuries across

Great music and brilliant summer skies – fantastic festival memories.

Michaela and I enjoy a backstage view of our festival.

Meerkats – one day I'll understand what goes on in their heads.

Dustbath – has any donkey ever had this much fun?

Arthur and Alice – height difference needn't be a barrier to ungulate love.

Zebras – an unexpected head-turner for passengers on the Norwich to London train!

Blossoming family.

Chewing the cud with a Red Poll.

Norfolk, Suffolk and Essex, they were invaluable for ploughing and pulling. Put simply, they were the engines of the land.

'Punch' is an Old English word denoting someone short and stout. No one would mistake the Suffolk Punch for a Shetland pony – they're almost six feet tall from ground to shoulder blade – but they're very definitely a power pack. That immense strength and stamina would eventually prove their downfall. Thousands were dispatched to the Front in the First World War to pull the heavy artillery around. Few would ever make it back. The vast majority perished in the mud. In the years following, the Suffolk Punch was further marginalised by the mechanisation of farming, very nearly dying out, just a handful of breeding mares continuing the line. Personally, I think they're the most tremendous of horses; just to be near them is awesome. I look at them sometimes and imagine myself being up there in a suit of armour. The question is, what use they can have today. Having them here to be admired is one thing, but if we can actually use them in a working environment, even just for conservation grazing, then that would be magnificent.

I do have a rather bizarre family connection with heavy horses – my grandmother was a rag-and-bone lady. She had a horse and cart, and, like old man Steptoe from the BBC sitcom, wasn't averse to the odd cunning ruse. Her favourite was to see a kid in the street, tell them to go inside and bring out their dad's Sunday best, and then give them a little chick or a goldfish by way of return. A couple of days later, their irate dad would find out and have to go down to my grandma's yard to buy it back. She was also notorious for running red lights. An amber gambler on four legs.

No matter their magnificence, the Suffolk Punches will always be rivalled for popularity by the donkeys. Peggy, in particular, is edging towards her winter years. She recently turned twenty-one, which is knocking on a bit in donkey terms. She doesn't want to go out partying on her birthday, although we did make her a cake, but would rather stay in for a groom and massage, as would many of us once we've reached a certain age. Donkeys moult and so a wash and groom can really help them maintain a healthy coat. Sometimes I find myself taking a step back and saying to myself, 'Hang on! Am I running an animal park here or some kind of weird donkey spa?'

Having said that, I suspect few people would see a visit from the dentist as being part of a luxury experience. Wielding a large file, he sets to work taking a level off our domesticated donkeys' teeth. They eat softer food here than they would otherwise and so their teeth don't wear down as they should. Whatever the animal, it's important that we cater for elements that might be missing in their new environment.

Having excellent and highly knowledgeable veterinary help is huge. Unlike small animal practices that treat household pets, like any farm, we rely on those vets specialising in the larger beasts, the cows, pigs and horses of this world. Except of course we don't just stop there. As much as we might want a vet to check out a cow, occasionally we might also want them to do a pregnancy test on a tapir or a well-being check on an unusually quiet macaque.

In many ways, we've grown with our vets. The accrual of specialist knowledge has been a two-way street. Animals aren't like us. They can't nip down to the doctors, say they feel rough, have got a terrible headache, and point to exactly wherever it hurts. Take a bearded dragon to a vet and they might as well be looking at a log. It's not sitting there telling them, 'Well, doc, the thing is, it really hurts down the back of my neck first thing, but then something weird happens and by mid-afternoon I feel it right down in the tip of the tail.' There's no immediate communication to work from. They are basically animal detectives, trying to work out what the issue might be.

I truly do think vets are amazing. Whereas a doctor has a solid base of human biology to work from, wildlife park vets are investigating everything from fish to monkeys, wallabies to donkeys. Just look at the respiratory system. Birds have air sacs, spiders have book lungs, insects have spiracles, and monkeys have lungs. An octopus (not that we have one – yet) has three hearts! Imagine on a daily basis dealing with all that bizarre anatomy. It's like being the resident doctor in the Mos Eisley cantina in *Star Wars* or a GP on the USS *Enterprise*. One minute you're helping Scotty get over a virus, the next Spock's in the waiting room oozing green blood (Vulcans have copper in their blood as compared to humans' iron – I know this stuff).

I can bring a bit of veterinary know-how to the table. The cycle of parasites is something that fascinates me, and the knowledge I accrued at university has been particularly helpful as someone who keeps livestock. Understanding parasitic cycles is really important if you want farm animals to be fit and healthy.

I've also never forgotten doing a work experience placement at a vet near school. One day, they actually had a monkey brought into the practice. I recognised it immediately. At Mole Hall, we had cotton-top tamarins, distinctive for a gonk-like shock of white hair on the top of its head. The vet, however, declared it to be a marmoset. The two are similar, except the common marmoset has tufts of hair coming out the side of its head.

'It's a tamarin,' I told him.

He wasn't having it. He just looked at me – 'It's a marmoset.'

I gave up. I knew I was right.

Stopping short of having my teeth filed, one thing I've always done in summer is take time simply to sit down and unwind. There's something really sacred about a long, perfect summer's day. Its beauty, its generosity, is something I've long appreciated. At Mole Hall, I'd work in the wildlife park during the day before heading across to the farm to bale and stack the hay. Eventually, with the sun sinking in the background, it would be time to stop and reach for that first cold beer of the day. OK, I'll admit, as I got older, there was the odd occasion where I might not always have waited for the end of the day. Mrs Johnstone had a swimming pool which she'd turned into an outsized koi carp pond. One day, she asked me and the gardener to remove all the fish and put them in a moat. The only way I could conceive achieving such a feat was by using a fishing rod. So for the next few days, which just happened to be hot and sunny, we both sat there for hours on end, fishing while drinking Stella Artois out of teacups. For some reason, the number we caught tended to decrease as the day wore on.

While drinking Stella Artois out of teacups might be a thing of the past, as the shadows lengthen, Michaela and I will take the chance to head down to the pond and just have that one cold beer. An intensely beautiful summer moment. Everyone, except us, has their best outfits on. Damselflies zigzag in soft electric blue and green, green willows cascade to the floor, a chestnut kestrel reveals its speckled breast as it hangs in the air.

Meanwhile, the black swans, found predominantly Down Under, although colonies have set up in Buckinghamshire and Dorset, are starting to breed. Romantics love black swans. They mate for life, split the responsibility of looking after the kids, and have an exceptionally low divorce rate. Really they should be involved in marriage guidance counselling.

They share the water with the pelicans, another big favourite with children. That vast throat pouch, used to sift the contents of a great scoop from lake or river, the pelican filtering out the fish and expelling the excess water out the side of their beak, will always captivate. They are the most incredible hunters, eating anything they can fit in that beak, not just fish but birds, frogs and rodents. Sounds daft, I know, but I've always preferred pelicans to those other wildlife park favourites, penguins and flamingos. The latter especially look like they wouldn't so much as give you the time of day, inherently snooty, the avian version of Margo from *The Good Life*. The pelican, on the other hand, seems the mirror opposite, a good down-to-earth type – up for a pint of beer and a couple of good yarns down the pub. So long as they can fill their beak, they're happy. Incredible to think that pelicans have been around for at least 30 million years. Sad also to know that, having survived from prehistory, their numbers are now suffering due to water pollution and habitat destruction.

In summer, there are so many stunning spots to sit and contemplate, but my favourite has to be the orchard, transformed by Michaela from the undergrowth that faced us when we first moved here. To create somewhere suitable for farm animals was one thing. To bring colour and beauty to it, as Michaela strived so hard to do, was little short of a miracle, but then she did exactly the same with the garden, be it growing flowers, vegetables or herbs, so it's hardly a surprise.

From the blossom of the spring, the apples and plums will now be forming up and beginning to ripen, a hint of the bounty to come towards the end of the season. On the hottest days, however, the gift those trees give is immediate – shade. The relief of stepping from a seriously hot summer's day into any woodland is huge. The temperature difference is so marked it feels like taking a cold shower with your clothes on.

Other times, I'll take a stroll up to the pigs. I've no idea how many hours of my life I've spent just leaning on a gate watching pigs, but I do know I wouldn't consider a single one of them wasted. I find pigs endlessly fascinating, and at the same time hugely relaxing.

Don't ever underestimate the restorative power of the pig. In the past, we've had young people come to us from an organisation, based in Italy, which helps those with addiction issues. Over on the continent, they spend months on a huge farm together, be it working with livestock or making dairy and meat products such as salami. It separates them from the source of their problems and gives them a window on a different kind of life. They come here and lose themselves in the work, occasionally turning a few heads as well! Even Italian pigmen are pretty stylish! I remember one of them had a particularly eye-catching combo of amazing yellow jacket and wide-brimmed hat. It looked like he'd just stepped off the catwalk. We really have had it over here when even Italy's pig farmers put us to shame!

Modern pig breeds do have to be careful in the sun. In the past, farmers selectively bred pigs to be black or spotty to protect them. Their big ears, meanwhile, doubled as sunglasses. However, aesthetics changed all that. People didn't want black hair on their rind and so lighter pigs were favoured. While some rare breeds have thick bristles which offer some protection, they do find themselves, just like our albino wallabies, open to sunburn. If you've ever been badly sunburnt, you will know just how painful and uncomfortable it is, and it's no different for pigs. Renowned for their intelligence, pigs counter the threat of sunburn by covering themselves in mud – one of the best natural sunscreens out there (try it next time you're out on a ramble on a hot day), and cheaper than us rubbing factor 50 sun cream all over them. We created a wallow so they can cake themselves in a lovely cooling gloop. I look at them and can't help thinking of the very well-fed clientele of a beauty spa. The water cools them as it evaporates, while the mud also offers respite from flies. It's lovely to watch a pig wallow, one of those real treats in life. You just have to look out for the shake of the head, mud flying off those big flappy ears in all directions. It's not just the pigs who appreciate the mud. Swifts and swallows collect it for building materials – again, a lovely thing to spot.

I recognise I might split opinion here, but if I'm really lucky on a lovely warm day, I'll see a grass snake sliding across a pond looking for frogs (for those of a slightly less snake-loving disposition, there's always the odd fish popping up to take a fly off the surface). Easy to forget sometimes that we have three snake species in the UK. The rarest, found on only a smattering of heathlands, is the smooth snake, a constrictor that restricts the breathing of its prey and feeds mainly on small lizards. More common is the wood-, heath- and moorland-dwelling adder, our only venomous snake, naturally quite a frightening prospect, but whose bite is in fact of little threat to people other than the very young or old. Finally, there's the aforementioned grass snake. The largest of the three, growing up to 150 cm, and a lover of wetland habitats – its diet includes fish and amphibians – the grass snake is the one most likely to be seen around the wildlife park. Occasionally, I leave bits of tin out and about as they like to slither beneath them. There's few things more satisfying than to return a few days later, lift the metal, and find one coiled up. Grass snakes are also attracted by piles of rotting manure and compost, perfect not only for hibernation but for the female to lay her eggs, numbering anything from ten to forty. Again, easy to wince when you think of snakes being in your garden, but they are actually a great addition to any ecosystem. It's just that we're part programmed, part brainwashed not to like them.

We also have three lizards in the UK – the common (or viviparous) lizard, the sand lizard, and, also here on the farm, the slowworm, often mistaken for a snake, but which is actually a legless lizard ('The Legless Lizard' – surely the perfect name for a pub).

The fact we have snakes on the farm is another great lever to fire the imagination of children. They might come along to see our boa constrictor, native of South America, and think that's amazing, but when I tell them, 'If you're lucky, you might see one of our viviparous lizards or a grass snake,' then you really do see the wonder in their eyes. It's tempting for kids to think that British wildlife is boring. On nature documentaries, it can seem like Britain is all hedgehogs and badgers, whereas watching a wildebeest being taken down by a lion in the Serengeti is a bit more exciting. Thing is, the same life and death dramas are being played out in a handful of soil or on a bird table in your back garden. We always celebrate the predators or the big animals, and it's lovely to do that, but then let's also celebrate that simple caterpillar on the oak tree, because without that you wouldn't have blue tits, and without blue tits you wouldn't have the sparrowhawk swooping down with that trademark explosion of feathers. We have buzzards, kestrels, sparrow-hawks, barn owls, tawny owls and little owls here, all of them incredibly impressive – see a barn owl swooping low across a meadow and it really does stay with you. I like to stimulate people's imaginations to appreciate what we have got rather than what we haven't. Those airborne predators, along with the stoats, weasels, foxes, badgers and hedgehogs, indicate the presence of a thriving food chain with lots of prey items – moths, grasshoppers, voles and mice to name but a few. That blossoming food chain is in itself an indicator of a healthy environment. In recent years, Suffolk has even seen a return of the red kite, a mark of the success of the reintroduction of the bird in the Midlands during the 1990s. Kites are the ultimate scavengers and recyclers. In Oxford, for example, they make short work of the takeaways left strewn around by students from the night before. Often they will follow the line of a motorway, feeding on roadkill.

I love the fact that so often our non-native animals can be the spark for a fascination and love of native wildlife. On a rainy, cold early summer's day, a child won't see many butterflies out and about – they'll be hanging upside down under a leaf shel-tering – but when they walk into the jungle-like arena of the butterfly house and see the delicate vibrancy of the blue morpho from South America, with its wingspan of up to eight inches and incredible iridescent blue, or even the owl butterfly, with an 'eye' emblazoned on its wing to ward off predators, they'll be blown away. Suddenly, they can't get enough of butterflies – which is great because we have ten different tropical species. I'll throw in a few remarkable butterfly facts. 'A butterfly tastes with its feet,' I'll tell them. I don't care who you are – someone tells you that and you want to know more.

On the way out, I'll say to them, 'Next time the sun's out, see how many butterflies you can spot on the way to school, in the park, or on your gran's buddleia.' When they see that red admiral, peacock, or painted lady, they'll be struck by how remarkable that is too. Sometimes to captivate a child, you need that stunning effect first. It worked on me for sure. I can still smell the deep scents of the flowers in that hot, steamy butterfly house at Mole Hall, the intensity of all those tropical shapes and colours that were so alien and yet so intoxicating. Straight away I had an overwhelming feeling – 'This is something I want to be part of.' At that point, you are giving a young person something that will stay with them whatever they go on to do in life. They don't have to be the next David Attenborough, but from that point on they will always carry that connection to nature. I cannot overestimate how important that is.

Summer is when the butterfly house is in full swing, when they start to choose mates, embarking on courtship rituals that are never less than fascinating. The postman butterfly – it gets its name from the repetition of its path from flower to flower every day – is one I always love to watch. The males don't do subtlety. Emerging earlier than the females, often they wait on the chrysalis of a female. When finally she emerges, immediately they'll mate with her. She then heads off to lay eggs while the male, his life's work done, then tends to die. It would never make it as a Mills & Boon story.

The postman is part of a group of butterflies called helico-niids, which feed on pollen as well as nectar. The more protein they extract from the pollen, the more eggs they produce. Tracking any butterfly as it flits around delicately selecting the exact plant on which to lay its eggs is fascinating, but the postman butterfly especially so. The female lays its egg on the passionflower, tasting the plant with her feet and then gently making a single deposit. Remarkably, in a space just a metre square, you can follow this life cycle happening.

By evening, the butterflies tend to have roosted up, clusters hanging, like drooping fruits, from beneath the plants. Every now and again, if the plants are being watered, a huge flurry of little butterflies will fill the air as they head off, disgruntled, to find a new roosting spot.

The plethora of butterflies I see around the farm resonate with me so much, a sure sign that we are doing something right with our environment, nurturing it, treating it with respect, and receiving our rewards. It's one of the reasons we don't use weedkiller. Not only do we not want to put chemicals on the ground but the resultant increased plant life encourages a lot more insect life, which in itself feeds the birds and so on. There are weedy patches and rough areas all over to encourage that incredible summer hum of insect life. Any overgrown area

abounds with life, the sound of grasshoppers especially transporting me back through the years until I'm in the woods or
on holiday as a kid. An assault on the senses – truly stunning.
Along with the abundance of wildflowers – it's easy to forget
how rich our countryside can be if we allow it to thrive.

I'm building on that wealth by reintroducing hives to the
farm. We've had them in the past, an excellent and hugely
knowledgeable chap called Barry coming over occasionally
to help us out. The first time Barry turned up, I thought his
old estate car was on fire. The smoke actually turned out to
be from the ancient puffer he uses to distract the bees. Barry
would open the boot and immediately I'd be hit not only by a
wall of smoke, but by thirty bees flying out randomly and a
smell like furniture polish. Beautifully spoken, I always thought
of him as being somehow related to J. R. Hartley – 'Oh, hello.
Do you happen to have a book called *Beekeeping* by . . .' He'd
evolved this exceedingly gentle manner for a reason – you can't
be hurrying around bees, you can't be flapping your arms and
shouting and screaming. Barry was so comfortable around the
bees that he didn't wear a veil. There was nothing he didn't
know about them or their behaviour. In a film, he'd be The Bee
Whisperer. But then he had his mischievous side. Occasionally
I'd feel a sting, look down and he'd be holding a bee against my
arm without me knowing. At that point, he'd laugh his head
off – 'That's very good for arthritis,' he'd tell me. I didn't have
arthritis.

I preferred it when he was talking about honey. He'd tell
me how it changes through the seasons, how the colours get

MAKING A BEE HOME

While we often think of bees as living and working communally in hives, many are actually solitary creatures. They still need a place to live though, and the good news is that with a little work we can easily tempt them to take up residence in our gardens.

A simple cluster of bamboo sticks, or indeed anything small and tubular, such as reeds, straw, even bits of old untreated cardboard, offer a ready-made home for bees. If using bamboo sticks, bind them together – use canes between 10 and 15 cm long, as bees like to tuck themselves away – and then hang them up or wedge them in a gap. Remember to plug one end of the tube so it mimics a hole.

You can always add a waterproof layer by cutting a plastic bottle in half and inserting your bamboo bundle inside. Either way, the main thing is you are providing an easily accessible option – chances are a passing bee will be moving in before you know it.

A terracotta plant pot and small tray, meanwhile, make a great home for bumblebees. Scoop out a shallow hole in the soil and place the tray at the bottom, then place the plant pot, *upside down*, on the tray. Before doing so, put some nesting material such as straw or dry grass into the plant pot. Fill in the hole and where the pot sticks out above the soil, place a slate on two stones to provide a sheltered entrance via the pot's drainage hole. Hopefully a female bumblebee will deem your handiwork the perfect place to set up a nest.

Bee numbers have been dropping in recent years. Pesticides have taken their toll, as has loss of habitat, so anything, however small it may seem, we can do to help our friends, integral to our own survival, is always worth a go.

Even with the most limited of space, anyone can keep bees. If your planting space is limited to a window box, you can combine growing herbs such as marjoram, thyme, sage, chives, mint and rosemary with helping your local bees. Believe me, it's well worth it. Not only are bees fascinating to watch but, naturally, they're highly efficient at pollinating anything in the surrounding area.

darker and the flavour changes according to the flora on which the bees are landing, how a bee that started on the early hawthorn flowers may well end up on those of the late ivy.

'Try this,' he'd urge me, 'from this season's chestnut blossom.' And the most wonderful taste would fill my mouth. Every time he came, Barry would take me off into this beautiful world of beekeeping. Then, at the end of his visit, he'd pour me a glass of mead, so strong that by the time his old estate rattled off down the lane, I'd feel like I was on a different planet.

I don't want to give the impression that it's only the big-hitters, all things bright and beautiful, that excite me in the insect world. Bumblebees and butterflies get a lot of attention, but ants and earwigs are equally important. In the insect lab at

Coventry University, I had a small cup of parasitic wasps that I'd collected, the ones that lay their eggs inside caterpillars and beetle larvae. I looked through the microscope and they were tiny, like a pinhead, but at the same time an incredible mix of red, blue and gold. Immediately it reminded me of an old pirate film, where they prise open a treasure chest and it's full of rubies, emeralds and sapphires. It's so easy to look at flies on a riverbank and think of them as mere gnats, but take a look up close and they are so much more than that, an incredible organism in their own right.

Spend just thirty seconds standing and watching a tiny insect and you'll see it in a different light. You may see a little fruit fly flit over to another and message it with its wings. They might even perform a mating dance. I really do urge you – stop and have a look. I do, and afterwards I feel a hundred times better, same as I do when walking through a meadow that's been left to itself – it's an incredible thing to be part of something so alive. With every footstep in long grass, you become part of a whirl of hopping, scurrying insects, from leafhoppers to crickets, meadow brown butterflies, and beyond. Ladybirds, the ultimate con artist, are everywhere. People think they're so cute when actually they're just beautifully dressed assassins – Al Capone in a dress – big chunky destroyers ripping to shreds every aphid unfortunate enough to be in their path. Honestly, give your kid one of those little ladybird umbrellas and you don't know what you've done – you're associating your child with a tyrant. Its larvae are no better. Clad in orange and black, they're like tigers. Very little survives to tell the tale.

The top of the farm, which has been pretty much left alone, has become a particular haven for grasshoppers and crickets, as well as the grass snakes and lizards. From up there, you don't have to look far to see the contrast with the arable desert that makes up so much of the farming landscape around here – farms that are basically wheat-growing machines. At Pannington Hall Farm, I'm glad to be surrounded by so much of what traditional countryside should be. Immerse yourself in this miniature domain where insect life is rampant and suddenly the bigger world disappears. It doesn't hurt occasionally to see the planet in a different way. Those tiny insects you're staring at are actually the building blocks of life. In our human view, the smaller something is, the less significance it has, when actually it's the smaller things that are of the most importance. It's often said that there are more living organisms in a single teaspoonful of soil than there are human beings on the planet. Considering the human population of Earth is almost 7.5 billion, that really does make you sit up and think. Far from being the leaders of the pack, we're actually just hanging on to the coat-tails of the natural world. The insects are the most successful group of organisms in the history of the planet. Their evolutionary brilliance has led them through millions upon millions of years.

I see it as like a film. The big stars garner all the attention but none of them could operate if it wasn't for all the others, the lighting technicians, the wardrobe and make-up teams, the runners, supporting them. The natural world is exactly the same and we're always looking to do what we can to make sure

MAKING AN INSECT HOTEL

Why should people be the only ones to enjoy ready-made accommodation? How about creating a five-star, multi-storey, luxury insect hotel? It's what old pallets were made for. Cut them in half, stack them up and soon you've got a boutique stopover that's the insect equivalent of the highest-class Mayfair getaway. Well, if you like sleeping on a bed of straw, that is.

The reason pallets are perfect is they are full of gaps, into which you can push any dry material, dried grass, sticks, leaves – the basic garden detritus that is all around. Add to that the decaying wood, bark and cardboard beloved of the creepy-crawly community – woodlice, centipedes, beetles and the like – and you should soon have a wide variety of guests.

It's worth laying the ground floor on bricks. That way your hotel will also be open for larger visitors – small rodents, for example, and hedgehogs. You could even work in a few little cool-rooms, fashioned from a few strategically placed stones, for any passing amphibians.

Put a cover on the top, an old bin lid perhaps, maybe some roofing felt, and before you know it all sorts of beasties, from ladybirds to lacewings, will be invading every nook and cranny. It's not a totally one-sided act of generosity – when the guests check out in spring, they'll be straight into your garden, removing it of unwanted bugs and aphids. You've basically been playing host to your most useful predators. An insect hotel delivers great reward for minimal investment.

those links are recognised. While I'd be first through the doors, Termite World is unlikely to take off as a visitor experience, but that doesn't mean we can't raise the minibeasts' profile and show how they, and millions of other lesser-seen animals, play a vital role in maintaining the balance of the planet as a whole. The natural world doesn't begin and end with the animals so familiar to us. But we can get people interested in the tiny stuff, the animals that aren't always on our radar, by showing them the big stuff. What's great is if a child then walks through the woods and sees how important the insect life is to the wildlife on their doorstep.

That, for me, is the key to what we are trying to do here. Yes, I want everyone to marvel at the incredible array of animals we have, but I want them also to make the connection between the wildlife park and the outside world. This isn't Noah's ark – we don't have two of every species – but what we do have is a great illustration of how the biggest threat to the natural world is ourselves. So much of what people see and love here is either threatened by human activity or has ended up with us because of people's skewed attitude to wildlife's treatment.

Hiccup and Bobo, our North American raccoons, were rescued from lives as 'pets'. Kept in totally unsuitable conditions, they had been through a cycle of aestivation, a type of dormancy used to survive extreme circumstances such as excess heat or lack of food, and become very aggressive. Unsurprising really. Raccoons might have a look of a cartoon swagman with a 'mask' of black smeared across their eyes, but they're actually way more intelligent than anything you might find on Cartoon Network. To be denied anything even remotely resembling their basic living environment must have been torture for them.

Should any mistreated animal come to us, we'll do our utmost to make sure its living space is the best we can provide. Hiccup and Bobo had endured a tough life, kept in basic cage conditions, so we were keen to build them a habitat that reflected their love of hollow trees, rocks, crevices and ground suitable for making dens. We wanted them to be able to enjoy the incredible gifts nature has given them. Their paws, for example, are capable of sending three-dimensional images to their brain, basically allowing them to see in the dark, handy (no pun intended) not only because they're nocturnal but also when seeking a fishy treat in dark and murky water. We throw bags of food and fish in the pond in their enclosure so they can still enjoy that element of their wild selves. They also use that intelligence to become amazing escape artists. Like the pigs, they know their enclosure is surrounded by an electric fence, and equally they know that when the electricity goes down it's time to go exploring. I expect if you listened hard enough, you'd hear a little snigger as they crossed the wire and headed off into the park.

Possibly because of the way they've been treated, Hiccup and Bobo don't like to be handled. Unfortunate, because to feed them, check their health and clean their enclosure, we need to get up close. Raccoons have razor-sharp teeth and claws – an omnivorous diet means they have to tackle everything from nuts to rodents, frogs and crayfish. Bearing that in mind, we were probably lucky that the worst they could throw at us was leg-nibbling. We came up with the novel solution of taping thigh-high cardboard to our legs for armour while bribing them with their favourite treat of home-made popcorn. Initially when we went into their enclosure, they'd come after us. Drop a little bit of popcorn, though, and they'd instantly be distracted, which was handy because not being attacked while working at a wildlife park is quite important. The popcorn changed their opinion of us for the long term, and next time we visited, two weeks later, they'd put the leg-nibbling in the locker. Good news because thigh-high cardboard leg guards really aren't the most flexible. From the outside, it must have looked like Hiccup and Bobo were dealing with a zombie apocalypse.

We might have achieved it in a somewhat unorthodox manner, but that little game was all about building up trust. The key to training animals is little and often, something I take right back to a little budgie I took in as a child. This little bird was really flighty – every time I went near its cage it would flap around. At first, I accustomed it to simply being looked at by me. After a while, I opened the cage and held a little stick for it to hop onto. It took time, but eventually it did so. After a while, my finger became the perch instead of the stick. Over time, the bird came to trust me to the point where I could get it to hop on my finger, take it out of the cage, let it fly up to the curtain rail, put my finger out, watch it hop back on again, and put it back in its home.

Ultimately, the plan with our kookaburras is to train them to fly to the hand – that way people can see this fascinating, beautiful and slightly comical bird (look out for that knowing smile) up close – and essentially it will be the same training routine as with that boyhood budgie. Those who work with animals know how important it is to build trust. For all the times on TV you see cowboys herding cattle, galloping round, whooping, cracking whips, truth of the matter is it's much easier and more efficient to walk in front of cattle and get them to follow a bucket. It's just that it doesn't make for quite such exciting telly!

The coatimundi, which belongs to the same family as the raccoon, is another that all too often finds itself caged in someone's home. I love the coati, not just for its inquisitive face and big bushy tail but for the origin of its name, from the native South American Tupian Indian words 'cua' (belt) and 'tim' (nose). Confused? Well, coatis tuck their nose against their stomach when they sleep. We have five, Coco and her son and three daughters. Wanting our coatis to feel at home, we built them a wooden shelter. They thanked us by taking themselves off to build their own pad in a tree, preferring to use our lovingly erected shelter as a toilet.

I feel utterly bemused when I hear people are keeping such wonderful wildlife as pets. It's hard to imagine the mental and physical stress a coati or a raccoon must suffer. It's why providing a new home for such animals is so important to us – a halfway house between rescue and conservation. And a reminder to us all that the natural world does not exist to give us something to look at on TV once in a while.

The list of wildlife caught in this constant cycle of abuse is, sadly, endless. We are often in contact with Heathrow Airport which has an ever-rolling conveyor belt of unlicensed, smuggled and mistreated animals coming through, ranging from pumas to lynx, serval cats, lizards, turtles, tortoises and monkeys, all of which need rehoming. Like anyone, I find these stories hard to hear. On a basic level, I like the idea that finally we are giving animals which have suffered heart-breaking abuse at the hands of humanity a break.

While we will always step in if we believe we can help, most animals that come to us do so after a good amount of planning. Act too hastily and we can adversely affect the stability of the wildlife park. It's vital to consider the room we have and how any new arrival will fit in with our long-term collection plan. It's not as simple as just taking in any needy animal. They need the right facilities and people to look after them. Over time, it might be that we are not doing them or ourselves any favours.

These are the decisions being taken by wildlife parks and zoos the world over. A lot of people remain opposed to keeping animals in captivity, anger I totally understand when they see them neglected and stuck in cages in dingy, dirty and badly run establishments. However, most wildlife parks and zoos, like ourselves, treat animals with the greatest respect. With the decimation of natural habitats and animal populations, they also act for the common good. There's another overwhelming factor to be taken into account – these animals have nowhere else to go. It's not like we can just open the gates and let them all out. There's a philosophy against keeping animals in captivity, but there's the practicality of dealing with a problem.

As it stands, mankind is the greatest risk to global animal life, and wildlife parks and zoos are doing their best to redress the balance and ensure endangered species survive. While each has its own individual flavour, essentially we're all working together to build a wider genetic pool, vital to rebuild populations. The problem is that, while females can physically deliver the next generation, half the animals born in captivity are males, which raises the question of what to do with all these boys. The answer is for zoos and wildlife parks to act as one large organisation. That way, while some are breeding, others can feature male-only groups under the banner of education and conservation. Our zebras, Zebedee, Zoom and Thunder are males, so while we can't breed them, we can inform visitors about them and the importance of maintaining their natural habitat in the savanna, helped by the fact that zebras have an infinite appeal and fascination. Zebedee and Zoom came to us from the Lake District Wildlife Park after their father, being the dominant male, pushed them out of the herd. Thunder then made up the trio, joining us from Longleat. We wanted only the best for the newcomers and so set about building a 33-metre house with six bedrooms, two courtyards and four holding yards. It might sound like something better suited to the pages of Country Living, but in fact this layout allows our keepers the space and facilities to work with the zebras in a way that suits both them and the animals.

This grand design formed part of our 6.5-acre African exhibit and getting it finished before the park reopened after lockdown was one of our most hectic tasks. It was vital we got it done, not just for the zebras but for another exciting, new and ambitious addition – Barbary macaques, the first primates on the farm. It wasn't just hammering in posts. If a macaque is to be made to feel at home, it needs more than a fence, which meant an imaginative solution to the issue of housing – converted shipping containers. Spacious and easy to clean, it was hard to think of anything better. The plan came to fruition uncannily easily when a shipping company announced they'd be more than happy to help us out – one of the benefits of being near to the UK's busiest container port at Felixstowe – although we were left wondering whether it really was our finest brainwave when the wind got up on delivery day and we had two big lumps of metal swinging on the end of a crane.

From that point, we had just five weeks before our guests' arrival. As soon as the containers were set on their concrete bases, we set about fitting them out with bedrooms and chambers. I looked around at the finished product and, thinking back with misplaced nostalgia to the caravan I'd started out in, felt a little bit sad I wouldn't get to spend a couple of nights in there myself. Relaxation, though, was very much not part of the process. To take the enclosure to where it needed to be there was still plumbing and electricity to be installed, a kitchen and medical facility to be fitted for the staff, and landscaping to be done.

At the end of all this, in came Adeo, Sisi and Ndidi, all three of them having been dealt a terrible hand in life. Adeo was a beach monkey, kept in a tiny box, removed only to be pictured with paying holidaymakers in Spanish coastal resorts. Sisi and Ndidi, meanwhile, were lab monkeys used for medical research. All three came to us from a rescue charity in the Netherlands which improves the lives of many animals – lions, tigers, marmosets and serval cats included – illegally traded or misused for the tourism and entertainment industries across Europe. The threat to Barbary macaques – the only macaque species found in Africa, and the only monkey species in Europe – is, however, on a scale far more widespread than that. In the natural world, they face a huge threat from widespread habitat loss and that seemingly never-ending desire to keep wild animals as pets.

Instantly they loved their new container homes, but what all of us really wanted to see was them go outside into their enclosure for the first time. Considering the life they'd had, it was an emotional moment to see them enjoy such freedom.

Adeo especially is an interesting character. He has a penchant for throwing the odd stone in my direction, preferable to one or two of his other hobbies which, this being a family publication, I won't mention here, other than to say he can do one of them with one hand while sticking his middle finger up at me with the other. To be fair, his middle finger is permanently in that position after being broken in the past. To my amateur animal-psychologist eye, Adeo seems a bit confused, as if he wants to be friends but there's something stopping him. Hardly surprising considering his background, abused and kept in a tiny cage. He's extremely complex for all sorts of reasons, none of which are his fault.

Adeo's behaviour might adapt. Already we've seen him respond to the stimulus in and around his enclosure. Early in the year, he watched the ducks come in to feed on the pond and took a real interest in their ducklings hatching out. When at one point our old friend the heron flew down and ate one of them, it was obvious from his quiet demeanour that he got really upset. Other times, he climbs to the top of his tree to watch the trains going past. Ipswich to London continues to be his dream commute.

At the same time, he really does deserve the title of 'cheeky monkey'. He had a habit of sticking his hand through the wire and grabbing my brother when he was working around his enclosure. Then when Stan the electrician was doing his bit in there, Adeo was forever pulling his plugs out. Stan would traipse over, plug his drill back in, and then, after waiting for him to get back in position, Adeo would do the whole thing all over again.

Eventually, without the help of Adeo, we created the small matter of a thousand square metres of walkways, climbing frames, ladders, ropes, platforms (to rest and to groom), shelters and ponds. Even a hammock. That's where the macaques now hang out, and, considering their past, it's hard to think of three more deserving animals. But it's important to remember that, while people love to see them swinging and climbing, the macaques also spend a huge amount of time foraging. As tempting as it is to think that with primates you build upwards, actually, as with any animal, once you investigate further you soon realise their needs are a lot more complex. When we see a macaque running across a rope or springing up a piece of timber, we are only observing one of many intertwining behaviours. The key is to build a habitat that reflects that spread of instincts and activities. For us, it was vitally important they had lots of space to indulge in what they love doing on the ground, hence the sheer size of the enclosure.

The macaques fascinate us all and none of us can resist trying new ways to stimulate their brains. Thinking it would keep Adeo quiet for a while, I once gave him a big cardboard box full of sheep's wool. He ripped it open, had a good rummage through, and then gave me a really dirty look. It was obvious what he was thinking – 'Wool? What do I want with wool? Where's the banana?' Food does tend to keep them very occupied – unfortunate for the ducks who lay their eggs around the pond in their enclosure. For a macaque, a duck egg is cordon bleu. Only when the egg becomes a duckling do they become more concerned about its inhabitant.

After several months, we took seven more macaques from the charity, this time more of a breeding group, on the whole younger and more active than those we already had here. Macaque life is complex, with a whole load of social conventions and hierarchies, and so just throwing the two groups together would be a recipe for violent disaster. There's an older girl with the new arrivals who would probably mix with the existing trio no bother, but there are also two dominant males. Put them in with Adeo and one of two things will happen: Adeo will think he has to terrorise the newcomers to dominate them, or, and this is the more likely outcome, the two dominant males will beat merry hell out of him. At that point, he'd either have to submit or go it alone and end up being totally isolated. It's worth neither a punch-up nor Adeo ending up on his own. Equally, the situation is likely to spark the kind of unrest which could result in real damage to the enclosure. There can still be interaction because they are able to see and hear each other through the fence. In fact, the first time his new neighbours came outside, so shocked was Adeo that he actually did that great monkey expression where their bottom jaw drops so far it almost hits the floor. Possibly this was because he saw the five females first. He might well have been disappointed to see two boys in there too. Nevertheless, both groups have been chatting away amicably enough (I think – I don't speak macaque). For Adeo, and me, it's been a nice change to him slinging insults at me when I walk past.

When they arrived, a handler informed us that one of the macaques is a prolific and accomplished escape artist. He was also very well travelled, having lived in four different countries. It made him sound like an international jewel thief, or a spy from an old black-and-white film. The good news is that, apparently, he escapes for no other reason than it amuses him.

He sits outside his enclosure and then, when the wildlife park staff come along to apprehend him, he jumps back in again.

For everyone at the park, the arrival of the macaques, as with the camels, was a truly pivotal moment. What we had here felt more than ever like a wildlife park. As usual, I couldn't help my mind wandering – 'As a kid, if I'd thought for one minute I'd have an enclosure for monkeys, I'd have never believed it.' And yet here we were. Almost twenty years on from when we'd started, we'd grown into a park truly committed to protecting wild species.

The macaques could have been seen as a finishing line. Instead they felt like the start of something big. We've since been adding other African species, starting with the eland, the world's largest antelope, known not only for its size and its tightly spiralling horns but – and this might be the most pertinent characteristic from our point of view – its ability to jump a four-foot fence from a standing start! Good job the area is securely enclosed. We'll build up other inhabitants bit by bit.

Not all our exotic animals turn up on the back of a lorry. Others make an entrance in a more traditional manner. Summer of 2020, for example, was marked by a very special arrival – a tapir calf for proud parents Tip Tap and Teddy. Tobias caused a real stir when he appeared, and I'm not surprised. A furry brown and cream bundle of stripes and spots, not only was he ridiculously pretty but in the Amazon his species is increasingly vulnerable. Loss of territory has found the tapir contained in increasingly shrinking areas, reducing its ability to roam and meet other tapirs with which to breed. Talking of which, tapir conception can be a shock to a passer-by. When aroused, a tapir organ can reach 19 inches. You really do feel their pain if, as does happen, they accidentally stand on it. The tapir isn't showing off. Its penis has to be that long to reach the female's fairly inaccessible genital tract. Should this union happen underwater, the penis also has a watertight seal near the end. You're right – nature really is remarkable.

A tapir's gestation period is thirteen months, but Tip Tap's timing couldn't have been better – the day after Tobias was born, we reopened to a post-lockdown world. What better way to celebrate? In all honesty, we hadn't planned for the pair to start breeding so quickly, preferring to wait a couple of years. Tip Tap was actually on contraception, but, well, accidents happen. On one occasion, we were petrified that a castrated bull with one testicle might have somehow achieved the impossible and impregnated some of our rare-breed cows. Only after a visit by a vet with a rubber glove did we stop sweating.

All our animals are ambassadors for their species, educators revealing the importance of maintaining the delicate balance that is the natural world, but of them all, the tapir is perhaps the perfect example. If a child is standing open-mouthed at the sight of such an incredible and outwardly unusual animal, then we can build on that wonderment by explaining how important a role it plays in the Amazon, an environmental engineer reinforcing those great rainforests by eating the fruit of the trees and spreading the seed via the vast network of waterways. We can talk about the Amazon being the lungs of the world and how important it is to preserve that habitat for the future of both its wildlife and ourselves. It's incredible how from a single species can emerge so many interconnected themes. At every stage, we want to use those stories to reach into and fire the imagination of young and old alike. It seems sometimes that when people reach a certain age they put that love of animals and nature away for fear it might seem childish. But an understanding of the natural world is a wonderful thing that, like reading and writing, should always be a part of us. An amazement at the natural world shouldn't come with an age limit. Children are often said to be mad about animals, but think about it – whatever our age, it would be mad not to be mad about animals. If we aren't mad about animals then we're separated from nature, and when that happens there's a problem. The more we fail to see its value, the poorer we become. Its degradation becomes a serious threat to us. Of all the societies and cultures I have been lucky enough to encounter, those that are tied into and deeply appreciate the natural world are always the happiest. They're the healthiest too, because that connection extends to their food. See the natural world as a conveyor belt serving our every whim and we become cut off from reality. That message can be presented in a preachy way, or it can be presented, almost without noticing, in the form of a child captivated by South America's largest land mammal.

We're extremely proud of our tapirs. They're our first European Endangered Species Programme (EEP) animals, part of an inter-zoo breeding scheme to keep bloodlines as diverse as possible, which in itself means a positive future for the species. The scheme also works hard to preserve natural habitats, particularly important for an Amazonian like the tapir.

On a personal level, tapirs are one of my favourites in the park. They are incredibly peaceful and, sliding in and out of their water, seem to find a natural joy in simply existing. With that huge snout, their design looks somewhat haphazard, but actually they have developed over millions of years to be supremely suited to their rainforest environment. Their body, for instance, is shaped like a teardrop, narrow at the front, broader at the rear, to enable them to penetrate thick undergrowth. Small ears and eyes mean less chance of injury. The nose, meanwhile, is essentially a mini-trunk. Like an elephant, they can pick vegetation and place it in their mouth.

Tapirs are also very docile and affectionate, particularly partial to a tickle underneath the chin, not to mention a banana. So attached to this treat are they that they greet the sight of their keepers with squeaks and whistles. Always nice to feel wanted! That said, their patience was tested when, over time, the water in the small lake we'd dug out for them slowly drained away. A few muddy puddles is not what they signed up for and we had to get the mechanical digger out to properly dam and reinstate the area before refilling it and putting a smile back on their faces.

Visitors are often surprised that our capybaras also spend so much of their time in the water. The world's largest rodent, outwardly they look like nothing more than giant guinea pigs, and anyone who kept one of those as a kid will, I expect, have spent very little time with it in the bath. Look closely at the capybara's feet, however, and you will see it has webbed toes. Like the hippo, meanwhile, its ears, eyes and nostrils are found towards the top of its head, meaning it's able to stay submerged in water with very little of its body showing, helping it avoid detection by predators such as jaguars and anacondas. Capybaras can even have a nap in water, keeping just that familiar flat nose above the surface.

The capybara is native to the Amazon basin and, as with so many South American animals, its habitat is increasingly under threat. It is also poached for its skin, in demand for leather. At the wildlife park, however, they are looked upon with nothing other than total fondness. Naturally quiet and friendly creatures, caught in a permanent smile and blessed with the look of an outsized pet, it's unsurprising that children in particular are so taken with them. Along with the meerkats, if there was one animal your average infant could sneak into the boot of their parents' car, I reckon this would be it. However, they are barrel-shaped for a reason. Eating half a stone of vegetation a day, and equipped with sharp and sizeable teeth that never stop growing, they move around like mini-bulldozers. No two ways about it, they'd make short work of the average garden, although your lawn would get a break for a while in the morning. That's when, like rabbits, they eat their own poo, delivering the bacteria which helps to break down the endless intake of fibre.

Our collection of vast rodents doesn't begin and end with capybaras. We also have a smattering of Patagonian mara, the world's third largest rodent (the beaver takes second place), which looks a little like a halfway house between a hare and a capybara, its incredibly powerful back legs giving it the ability to hop, walk and do a particularly impressive bouncy run.

Barren patches on the trunks of the trees are the giveaway that the mara share their enclosure with another of our great characters, Basil the giant anteater. He likes nothing better than to search for bugs beneath the bark. I often describe the anteater as an animal that looks like it's stepped straight out of a kid's drawing – if that child had been asked to make up an animal that could never exist. Like the duck-billed platypus, the anteater looks like it was put together by committee. 'Do we want claws?' 'How about an extraordinarily long snout?' 'What does everyone think about a massive tail?' Most remarkable of all Basil's attributes, however, has to be his spaghetti-thin tongue which comes in at a mindboggling 60cm. Smeared in sticky saliva and covered in tiny spines, the anteater darts its tongue into anthills and termite mounds. With the added penetration of its elongated head, it swiftly and efficiently secures a feast of hundreds of insects. After around a minute, during which it will have fired its tongue into the mound up to 150 times, it then moves on – ants don't take an invasion of their home lying down and will bite and sting to repel the intruder.

Anteaters have poor eyesight and so a heightened sense of smell helps them search out their next meal. Fortunately, in the forests and grasslands of Central and South America, anthills and termite mounds are abundant. We're a little lacking in both in East Anglia, but do our best to mimic Basil's natural feeding style by filling plastic bottles with baby mealworms and beetles. That way he can stick his tongue in and root them all out.

We might be unable to replicate his natural feeding environment, but Basil still very much has the serious set of claws he would otherwise use to rip open the hardened exterior of a mound. In captivity terms, he is, therefore, defined as a category one dangerous animal, the same classification as lions, tigers and bears. Rightly so – his claws could easily disembowel a human. That's not to say Basil's in any way a violent chap, he's actually got a lovely gentle nature, it's just that he's permanently equipped with an offensive weapon, even walking on his knuckles to keep his claws sharp. I prefer to concentrate on something rather gentler, his big, brush-like tail, which doubles as a blanket. Essentially Basil's a massive shrew, constantly having to eat high-energy foods and keep himself warm. Fortunately, he's perfectly equipped to do both.

One day, I envisage the tapir, capybara and mara all living together with Basil in one big South American enclosure. There'd be the pond for them to swim in, squirrel monkeys in the trees – perfect. They could all sit around and chew the fat together.

The idea now is to find Basil a female companion. Numbers of giant anteaters in the wild are rapidly declining – this is an animal that has been on Earth for 25 million years – due to habitat loss, wildfires and hunting. As ever, the work of wildlife parks in maintaining the species is vital. Basil's female friend will have to tolerate some of his more unusual behaviour. We found early on that he likes his home comforts, perking up, like Blaze all those years before him, at the sound of the radio. Also like Blaze, he's a fan of *The Archers*. I tell you, the long-running soap's audience figures wouldn't be nearly so healthy without the listenership we provide here.

Thinking about it, when finally we do up the old farmhouse and turn it, as planned, into a little guest house, maybe Basil would like to hang around in the kitchen, spending all day listening to Radio 4. After all, our idea is to make those who stay feel like they're in the middle of an enclosure, a place where animals populate the grounds rather than people. That way, when they stroll onto the balcony with their pink gin, they'll feel so among it all, like I was when I used to sit on my bedroom floor surrounded by all my weird plants with jungle sounds playing. Difference was back then that I had a pink lemonade.

At present, while guests would indeed be surrounded by occupants of distant climes, any illusion of being on another continent could well be shattered by the sight of endless bouncing white bunny tails. There are some sandy areas on the farm, the walkways around our new enclosure being a case in point, which make for perfect burrowing terrain. The presence of so much sand seemed a little odd until we started finding seashells and realised the estuary of the nearby River Orwell would once, many hundreds of years ago, have passed through here. It's not just the rabbits who appreciate the sand, so do the nesting bumblebees, and the incredible sand martins which so often mesmerise with their acrobatics.

There are, though, some animals which will always transport me to a different place. Catch sight of a peacock flying up into a tree, hear its trumpet, and straight away I'm in India, this magical bird's natural home and where it's considered so culturally significant. Another time I'll hear the kookaburras and find myself in the Australian Outback where, in the eucalyptus forests, the carnivore feeds on anything from mice to reptiles and snakes. That cackle also instantly reminds me of the old *Tarzan* films I watched growing up. 'It's Africa, not Australia!' I'd shout as, inevitably, the kookaburra sound was cranked up in the background. I'm a bit of a geek like that. I've lost count of the TV shows I've seen where a South American monkey has been passed off as African. I need to stop getting annoyed about it really!

Mercifully, the summer is perfect for shaking off headaches, major or minor. Sunsets across the park are rarely less than stunning. I stare into that incredible blend, crimson to yellow, and think, 'No way can it get better than that,' like a firework display, but of course it does! As if the sky is ablaze. And then it's gone; that very fleetingness is what makes it so special. As much as we like to record every special moment of our modern lives, you can never quite capture a sunset. A photo somehow never does it justice. To truly be savoured, a sunset must unfold in front of your eyes. We're lucky to have that in Britain. I've travelled all across the world, but I really do think our sunsets are unbeatable. In the tropics, for instance, darkness comes as if someone has flicked off the light. They say if you can put three fingers between the horizon and the sun, you've got thirty minutes – ten minutes per finger.

With the sun over the horizon, we are left in a magical crepuscular environment, a twilight in-between world before darkness falls, although high-summer days are so long there are times when it feels like the light will never truly fade. Now the moths start coming out. Butterflies, in all their finery, grab all the attention, but moths are hugely important for pollination. More than 2,500 species live in the UK but overall numbers have decreased by almost a third in the past fifty years, again due to intensive farming, a knock-on effect being less food for another twilight lover, the bats which flit across our park, and also the many birds which feed moth caterpillars to their young.

Animals don't tend to do lie-ins. When the sun's up, so are they, and right from the get-go there's work to be done. Glad to say, while there was a time when barely had darkness fallen than I'd be up with the dawn chorus, woken either by my alarm or a particularly vocal cockerel, today, with thirty full-time staff (and up to ninety staff of all kinds in high season), I'm no longer on conversational terms with the lark. Even so, while I might not be up at 4 a.m. these days – well, not often – I've never been one for lying in bed. In fact, I've often thought I'd be a good wildlife park exhibit. I have a problem keeping still, wake early, and from the off am up and about being busy. No one would ever be disappointed I was asleep in my shelter. I suppose the difference between me and the other animals is that I roam the entire park.

I've always liked how summer offers an extended window, and in the early morning, that lovely cool time before the sun starts blazing, it feels like anything can be done. It's like doing two days' work in one – a useful mental booster, especially at those times when the list of jobs is longer than your arm.

For some, though, long sunny days are a bit much. As Bactrian camels, Alice and Arthur are wired to grow a thick coat to keep them cosy in the intense winter cold of the central Asian steppe. Midsummer near Ipswich and they're wishing they had something a little lighter in the wardrobe. Instead they must wait while about 5lb of hair falls off in clumps. In human terms, camels always look an absolute shambles when they're shedding. Think of the scruffiest person you've ever known and multiply by ten. It can't be pleasant to wander round with great lumps of yourself hanging off and so we do what we can to help them out by mounting a broom head on a carousel so they can rub against it and speed the process up.

Our sheep too carry a huge fleece around through winter and spring. Now, as the temperature shoots up, it's the last thing they need. Early summer is when they're shorn, not something I've ever been involved in myself – all that mauling and manoeuvring in the heat is such hard work. We have a shearer who comes in and does the business with the clippers. Inwardly, I always feel the sheep's relief when finally that big layer of wool comes off. I'm always reminded of wearing a thick duffle coat when I was a kid, with a jumper, shirt, vest and God knows what else on underneath. It was like wearing a radiator. I can still feel the joy of taking that coat off, like being released from a 13-tog straitjacket. But then of course the sheep can't win. Post-shearing, they look ridiculous, like an army recruit who, in two minutes, has gone from flowing shoulder-length hair to a buzz cut.

Years ago, it was said that the money from a good supply of fleeces would pay the rent on a farm. Now, they're virtually worthless. So sad, because wool is such an amazing environmentally friendly material, not just for clothing but also for insulation and as a weed suppressant (lay it down, open up some holes, and plant through it). We make no effort to sell our fleeces. We keep them to show kids or leave bits around the farm for birds to take for nest boxes.

Of course, while in memory, especially childhood memory, every day of summer is hot and dry, the reality does tend to be a little different. I'm lucky, though, in that the smell of rain on dry earth is something else that spins me off into a world of imagination, a reminder of times and places. Watching a big summer raindrop throwing up a tiny puff of dust as it hits compacted soil makes me feel like I'm in the back garden aged five years old. Or jumping into the caravan to escape a rare unexpected downpour during that first summer here. I heard recently the name for this phenomenon of rain and earth combining to create its own strange perfume – petrichor – the root of which is Greek. '*Petra*' means stone, while '*ichor*', in Greek mythology, is the golden fluid that flows through the veins of the immortals. That might sound a little OTT but there is evidence that rainfall can revitalise. When it lands on grass, for instance, it acts as a catalyst for the plant to release extra oxygen. Maybe that's why when we're out in the rain we often feel more alive. Go outside after it's really thrown it down and the air is as fresh as anything you've ever experienced.

I'm not saying I jump for joy every time the heavens open – being soaked to the skin is nobody's idea of fun, and there's been many a time out on the farm when I've cursed the rain – but it is definitely the source of some of my favourite memories. I remember as a kid going bird-spotting with my old friend Simon. We found this big old metal water tank and shoved it up in a tree, even scavenging a bit of carpet to give our new base a touch of luxury. Hour after hour we sat there with our flasks in the hope of seeing a kingfisher. The rain softly pitter-pattering against the metal was the loveliest sound. A few decades on and the rain relaxes me in different ways. It gives me an excuse to switch off from the jobs that need doing outside and put my feet up to watch an old Ealing comedy for a couple of hours. Sorry, roof, you're going to have to wait – I've got a date with *The Ladykillers*.

As the summer drifts away, and I watch the goldfinches, grateful for a feast of thistle seeds as the natural bounty begins to dwindle, it's a moment to reflect that not only is autumn just around the corner but winter too is looming. It may seem strange to think about winter in summer but nature can be unforgiving and can punish you and, therefore, your animals if you don't make the right preparations early on. Fail to buy feed and bedding now and you'll regret it later. 'Make hay while the sun shines' – that's what they say, isn't it? But it's only when you go into farming that you really understand what it literally means – 'Make some hay now! Or in a few months' time you'll have had it!'

There's few things worse than running out of hay and straw in the middle of winter. Rest assured, everyone else will have got in there before you. What's left will be more expensive than printer ink – or Kouros. In the same way, now is the time to make sure vehicles are in good order and buildings are well maintained. When the roof of your shop blows off in January, you really do want your Land Rover to start first time as you pursue it across the fields. Over time, because it's part of the farming cycle, that kind of thing becomes second nature, but if you're not careful, they can be lessons learned the hard way.

There have been years where summer and autumn have been barely distinguishable, or in fact where an Indian summer has been rather better than the real thing. I'm all for a bit of late sunshine. The pigs might enjoy a roll in the mud but personally I'm not so keen. However, in many ways, I'm impatient for the switch of season. Spring and summer are times of gentleness (well, supposedly!) and there's part of me that hankers for the changes – the colours, the atmosphere, the slow drawing of the curtain – that autumn brings.

For me, the end of summer delivers not finality but the opening of a door on to a sepia-tinted world of wonder. It reminds me of darkness closing in as a kid, sitting reading wildlife books in my bedroom, opening the garage door to the light of various reptile tanks, walking through woods in a magical half-light.

I'm lucky still to have that link with my past, one which I can access so easily. It's there all around me. I don't even have to reach out to touch it. The seasons make it my past, my present, my total reality.

Autumn

Nature's slowing of the wheel,
a beautiful time to sit back and reflect.
Golden sunsets, leaves falling and animals
hunkering down for the winter ahead
– with a few notable exceptions.

Dolph is a reindeer. He's usually a fairly placid kind of chap – the sort who'd fit into any festive tale. Except in autumn. At that point, he's more like Casanova crossed with a bare-knuckle boxer. Pumped up to the max; no way could I go into his enclosure. Immensely strong, he'd pick me up with his antlers and throw me like a rag doll. The reason for this Incredible Hulk-esque shift in behaviour – the roaring, the posturing and, if necessary, the violence – is simple: the testosterone overload that is the rut. While those few weeks unfold, Dolph is classified right up there with our most dangerous animals, placed firmly in category one.

The shift in Dolph's personality is one of the most remarkable features of the year. In the wild, reindeer must literally fight for the right to breed. The sight of two bulls locking antlers, blind to anything other than ousting their competitor, is as astonishing as it is impressive. Make no mistake, antlers are an incredible weapon, virtually indestructible – in prehistory, people used them as digging tools, axes, all sorts.

While Dolph will drop his antlers after the rut, well before Christmas, our females, Mistletoe and Rowan, retain theirs until they calve in spring. That means we can say without risk of contradiction that the reindeer we see flashing across the sky on Christmas Eve are all female. Rudolph must be a girl's name! For pregnant females, antlers are a useful defence against predators as well as an aid to finding and defending food in the unforgiving Arctic and sub-Arctic territory which they inhabit. Antlers uncover food by doubling as shovels (in fact, caribou, as reindeer are called in North America, actually means 'snow shoveller' in French). That constant search for nutrition means reindeer can easily cover 3,000 miles in a year. Unsurprising then that they carry around a change of footwear for different conditions. Hooves expand and splay in the warmer spring and summer weather to give better grip on the soft ground, while shrinking and tightening to deal with the rock-hard terrain of winter. Another incredible feature of their evolution, a unique ability to see ultra-violet light, helps the reindeer distinguish predators in the blankness of their winter environment, and potentially see nutrients beneath the snow. Of course, it provides little defence against shotgun-wielding people. Wild reindeer populations are decreasing as they are hunted for their headgear and fur.

While us humans feel the nights drawing in and want only to get comfortable by the fire, for Dolph, as with many animals, it's that very change in day length that triggers such a marked hormonal transformation. To watch Dolph in autumn is to behold an animal totally immersed in its surroundings. He has no other bulls to battle but so hardwired is his instinct that often he'll try to fight the sets of antlers we have on display around his enclosure. His only concern is to protect his ready-made harem of Mistletoe and Rowan, often joined by breeding females from other collections, strutting proudly up and down keeping a watchful eye on them.

By the time he drops his antlers, he's exhausted. In real Jekyll and Hyde style, it's like nothing has happened and he just wants to be best mates again. He'll come up and give me a little shove – or two, or three. He reminds me of that kid at school who's always there – 'All right? Come on! What are we doing now?' That annoying lad on the back seat of the bus chucking an empty Coke can at your head.

It's not just Dolph who we see undergo this remarkable change. The air is full of the noise of all the wild deer in the woods. Every year, there's an amazing migration of red deer stags on their way to a nearby estate with lots of hinds. Roe deer are also moving through, as are the fallow deer bucks after the does. Often I'll hear the latter clashing antlers – an amazing sound.

We don't see the results of that activity, but come spring, we do see the incredible splendour born from Dolph's autumnal transformation. Last year we had two beautiful calves, a female, Ivy, and male, Ryobi. They stayed with us for a few months before, in order to protect the bloodline, moving on to another wildlife park in Scotland. As tends to happen on these occasions, when the new owners came to collect, they brought a few animals the other way, two rare-breed billy goats, a Bagot and a Guernsey, and a magnificent black swan. The Bagot immediately became one of the rarest animals we have, with only an estimated 250 breeding females left. Numbers suffered due to them not being particularly useful for milk or meat compared to more popular breeds. Bagots can be a bit wild, and occasionally scream rather than bleat, but they have a beautiful black-and-white colouration. They are also excellent conservation grazers. We have had our successes breeding Bagots, with two sets of twins in recent years. There's few things cuter than a Bagot kid, that's for sure.

While billy goats' routine to attract the opposite sex is a little different to ours, sometimes covering themselves in their own urine to make themselves more attractive to the nannies, sheep are busy breeding in a more orthodox manner, ready for what we hope will be a good crop of rare-breed lambs in the spring. If that's to happen, then the ewes must be in top condition. Sheep are seasonal breeders. Miss the window and the chance has gone.

Norfolk Horns are tough hardy sheep that have been brought back from the edge of extinction. There was only one flock left after the First World War. Now there's about 1,500 breeding ewes, a nearby nature reserve giving us six to start our own flock. Of course, to produce lambs you need a ram. Thus we were introduced to Nero, a vast ram with even vaster testicles. Before we could put him in with the ewes, however, we would have to kit him out with a marking harness, so called because when the ram mounts the ewe it leaves a coloured mark on her back in a process known as raddling. I think 'love harness' would have been slightly more romantic, but then again who am I to interfere with generations of raddling know-how. Once fitted out, the ram is left in with the ewes for at least thirty days, time for them to go into heat twice. The result of all this work, which generally goes unseen by the public, is that come spring we welcome around thirty kids and lambs to the park. By autumn they're effectively teenagers. Unlike some teenagers, however, they aren't hugely keen to have their ears pierced, precisely what happens when they're given an identification tag, one of those jobs which, however many times you see it, always makes you wince.

Piglets arrive throughout the year. They're tough, like little bullets, but in those early days are vulnerable to being smothered if their mum inadvertently rolls on them. Alternatively, if they get pushed out of the way and are unable to feed, they soon get cold and lose strength. Every rare-breed piglet makes a difference. It's a tragedy to lose one and so we make regular checks to make sure all is going well. Not a bind from my point of view – each and every piglet seems to have a different, yet captivating, personality.

Many of our sows have been with us since they were babies, a bonus when it comes to birthing because they know and trust us, are used to us handling them, and allow us up close. With piglets, it's always good to get in there as quickly as possible to check out their health.

Often there'll be a runt, a smaller piglet which might not have received as much food in the womb. In the fight for a teat they may well suffer again, missing out on the colostrum crucial for survival. A runt may need taking out of the litter and nursing elsewhere, and thankfully they often survive.

Like our lambs, the piglets are ear-tagged. A mother pig who hears its piglet squealing is never going to be happy and so we separate them while the job's done. Thankfully, Mum soon forgets and doesn't hold a grudge, especially if a big pile of feed just so happens to turn up.

Thanks to our breeding programme, we now have a good number of Middle White pigs. Once known as 'the London porker' because of high demand for its pork in the capital, the growing desire for bacon from leaner pigs saw its demise. If we can keep producing Middle White litters, then that has ramifications for the breed way beyond our boundaries. If we have five females in a litter, then we can send the majority off elsewhere to breed. To build numbers further, we'll keep a couple and change our boar. That way we and other breeders are constantly cycling Middle Whites and building numbers, good news all round as its meat is increasingly back in fashion. Middle Whites are very well bred. If there could be such a thing as a posh pig, then they're it. It's very good of them to lower themselves to being around me. I expect most of them hope to be moved on to someone with better breeding.

Of course, there are some animals you really don't want to breed – because they'll never stop. Our meerkats, as mentioned earlier, are a case in point. There's only so many we can have in their enclosure and so the females are put on contraception – the fact there's contraception for meerkats being an eye-opener in itself. A pregnant meerkat is slightly different from the cute little creature so beloved by us all. She becomes aggressive as she seeks to banish other females from the group for about three weeks before giving birth. In a wildlife enclosure, being banished isn't easy. The last thing we need is more Steve McQueens making a bolt for freedom and a life where they're not victimised by a bad-tempered pack member with a birthing bump. Contrary to popular belief, I really do have better things to do than spend my life chasing meerkats.

Enjoying the autumn light is one of them. After the stark brightness of summer, the sun is no longer high overhead; its rays hit the fields and woodland at an oblique angle. Autumn sun doesn't blind, it doesn't burn, it cloaks. It brings an amazing life to a season too commonly thought of as one of demise and death. Autumn is actually one big party – the last hurrah of nature before the harsh realities of winter.

I love looking at autumn trees from a distance. Our woodland is beautiful all year round, but autumn is when it truly excels, like being at a concert and your favourite band coming back for an encore of all their greatest hits; the leaves turn from all shades of green to the most incredible russets, bronzes and reds. When the leaves start to turn, it's like a painter's palette – an artist playing tricks with colour and light. Just when you think you've seen something unsurpassably beautiful, the artist sweeps their brush across the canvas again and you're presented with a vision somehow even more stunning. Those colour changes come from the tree, as it enters dormancy, using a compound to seal the base of the leaf. The chemicals left inside then break down, revealing an incredible array of colours as they do so. Eventually, just the husk is left, which drops to the ground. As those leaves fall, straight away I go back to being a little kid walking home from school, kicking great swathes of them, or picking up armfuls and having leaf fights with mates. Other times, a storm will blow in and give the woodland an almighty refresh, stripping the trees and forming a carpet of fiery shades made treasure-like with raindrops. I know it's not for everyone, but I quite like a bit of unsettled weather. I enjoy that wet musty woodland smell. I know also that the occasional downpour is vital to aid decomposition, tiny invertebrates breaking down that rich woodland detritus. The decaying material smells of life, energy and vigour and shows that preparations for the spring rebirth are already well under way. As soon as that food becomes soluble, the trees can suck it up. Rain is the ultimate driver for the ecosystem.

Within that woodland cathedral of colour, so grand and serene, there is basically a mini-riot going on. In those final mild autumn days, it really does feel as though every animal is going like the clappers, feeding up as much as possible, making the most of this last great bounty as an incredible abundance – hazelnuts, acorns, beechnuts and sweet chestnuts included – starts to rain down from above. The smell of ivy is a huge reminder of that. One of the final plants to flower, this late guest at the party is fantastic for the bumblebees, particularly the young queens hibernating for the winter.

For thousands of creatures, like the squirrels hoarding nuts and acorns in a makeshift larder, or butterflies feeding on big fat blackberries as they split open, and ladybirds seeking warmth in numbers beneath the tree bark, this is the remaining chance to prepare for the challenges to come. All those animals know, through that incredible mixture of instinct and design, that while it might feel like autumn, with its gentle touch, will go on for ever, all of a sudden, like a snap of the fingers, the cold will be upon them.

Standing underneath a canopy of leaves is a great place to just stop, take in and appreciate what goes on in the natural world, be it fungus gnats dancing around or spiders weaving their intricate webs, but more than anything my real enjoyment of our woodland comes from watching other people gain so much pleasure from it. To engage with kids, we leave lots of big sticks and branches around so they can build dens, just the same as I used to do. So many children grow up in urban surroundings and never have the opportunity to mess around in the open air, to build things, to invent games in their head, to really explore their imagination. I myself wasn't averse to pretending I was Robin Hood, a stick as my sword, hanging out in my den awaiting an inevitable battle with the Sheriff of Nottingham.

What might look like a pile of wood to you and me might be a spaceship or a fortress to a kid. But there's more to it than that. What I also love about the woodland is that parents can come here and be kids too. Some of the structures both kids and adults construct are truly incredible, more like small buildings than dens – the sort of thing that generally requires planning permission! We've come up to the woods on occasions to be presented with a 'den' four stories high. It's like Bear Grylls has spent the night.

In another nod to my own childhood, there's conkers everywhere. My girls go out and collect bowlfuls of them and I might even string a few up for a fight, reminded instantly of being in the classroom, reaching into my pocket and finding two old conkers in there, or a mate claiming the best way to make a conker invincible is to soak it in vinegar, rub lemon juice on it, or bung it in the airing cupboard for a few days. A long-unbeaten conker took on mythical status – until the inevitable day when a new contender came along and smashed it to pieces in the playground.

We try to fire children's imaginations further by putting models of native woodland animals such as foxes near the paths to inspire them to think about what's out there. Wildlife is around us all the time – we just don't see it. When we're making noise, everything clears off. As soon as its quiet and we're gone, however, it comes alive. We have an active badger sett in the wood, home to another animal that kids find fascinating (especially when they find out badgers can eat hedgehogs!), and the knowledge they are walking where a badger may have been snuffling just hours before makes the woods ever more intriguing.

We also have two deer models, but they kept getting trashed. Eventually we put cameras up to see what was happening. It turned out to be not an act of vandalism but a male deer full of hormones mounting them and in so doing smashing them to pieces with their antlers.

We also want kids to think about the animals that would have lived here, and so we have representations of sabre-toothed tigers, wolves, bears, all sorts of bits and bobs they would never normally associate with woods in England. We even have a Sasquatch, the ape-like creature claimed to roam the forests of North America. Actually, what we really have is a Saus-quatch, i.e. a Sausage-squatch, a creature I invented with a backstory that it died out in 1983 amidst the great sausage migration of East Anglia. I must have been going through a particularly imaginative phase when I came up with that one.

If the Saus-quatch fails to scare them half to death, then surely the velociraptor will. Unlike the Saus-quatch, it's also historically accurate. The point of the model is to illustrate that exactly these sorts of dinosaurs were, millions of years ago when this was a tropical paradise, actually running around right here. That's a really big 'Wow!' moment for a lot of kids, a point where they really do start thinking about the development of the natural world on a grand scale. They see the velociraptor's feet and then look at those of our hens and immediately see the similarity. What outwardly might appear just a fibre-glass model is actually creating a really important link. We have a model of a cave lion too. The idea of lions being in Britain is incredible to children, but they were indeed here until around 11,000 years ago, fairly recent in the big scheme of things. Look hard enough and you'll find a mammoth – 14,000 years ago you might have bumped into one of those too. Bears were plentiful in Britain until the early medieval period, and actually, having spent a little time in the Yukon, that vast Canadian wilderness populated by brown bears and grizzlies, I'm not altogether upset at the fact they've gone. It's an odd feeling to go from being a person to an item on a menu.

Wolves are a little more recent in the history of British Isles extinctions. The last wolf was claimed to have been killed in 1680, although there are reports of sightings well into the nineteenth century.

Gaining children's interest is all about presenting little bits of insight without being too stuffy. We've always wanted the farm and wildlife park to be somewhere people can just wander, rather than be directed here, there and everywhere, bombarded with information and told to keep off the grass, and that's one of the reasons children want to come back again and again. I love seeing kids running around, playing in the woods, or whatever. I think we'd all be shocked how few children have that in their daily lives.

Our approach is all about allowing people to learn about their environment rather than preaching at them. If a kid comes to our farm who's never seen a cow, I'm not going to stand there for ten minutes and say, 'Oh, that's terrible! How on earth has that happened?' That's not going to make that child feel good at all. Instead, I'm going to say, 'Cows are cool! They're amazing to find out about! Come and have a look!' We're all the same – we feed off positivity.

We've always been aware of the difference accessibility can make to young people. While there's a lot of wealth in rural East Anglia, a liberal smattering of tweed jackets and mustard cords, Ipswich itself is a working-class town with more than its fair share of social issues, same as Felixstowe and Harwich up the road. There are a lot of low-income families in the area and we try to make sure that our pricing doesn't put us out of reach. I want coming here to be a real escape for those children, an absolute immersion in nature, so they can get up close and experience the size and presence of a Suffolk Punch, or stand and watch macaques move, run, explore and inter-act in as near as we can create to their natural environment. Who knows? Maybe doing so will spark a lifelong interest in conservation. I could be watching the next Jimmy Doherty – poor devil.

I have always felt that, whatever the subject, whatever their age, ostracising people, making them feel awkward or differ-ent by lecturing them about how something, in your opinion, should be, is the quickest and easiest way to change absolutely nothing. There doesn't always need to be two camps. Us and them. For and against. There doesn't need to be a divide. That attitude comes from my mum and dad showing me never to get too big for my boots. They would say to me, 'If there's someone sitting on their own at a party, go and talk to them so they feel comfortable and not left out.' It's obvious really – unless people feel included they won't listen to what you have to say. It's like telling someone, 'Spend more on your food and only buy organic.' Do that and you aren't considering either them or their situation. What if they can't afford your idea of what's good? They shouldn't be made to feel lesser in some way. Involving people is the way forward. Often there's a divide in agriculture – you're organic or you're not, you're vegan or you're not, and so it goes on and on. But nature's not black and white. There's always a middle ground where people can connect. Preach and you just cause alienation. I'm more about solutions than negativity. Connecting gaps, rather than driving a wedge between people who see things differently, is vital if change is to happen.

That's not to say we don't involve ourselves in debates. During the United Nations Climate Change Conference (COP26) at Glasgow in 2021, we tweeted on our Jimmy's Farm platform that 'the planet cannot afford 2°C'. We know that because we see through our own involvement in conservation work the devastation that climate change is causing to already endangered animals, be it through habitat destruction or shifts in human behaviour. Similarly, while we can't all be Greta Thunberg or David Attenborough, I do believe we can do our bit to try to live a good life both for ourselves and others, be it by recycling, reducing food waste, using less fuel, or whatever. For me, though, you'll never effect change by making people already juggling their own problems feel like they're being bashed over the head about issues over which they have so little control. It's not down to them to head to South America to stop logging and replant the rainforest, but they can be part of fun and positive ways to make a difference. Sourcing local food, planting shrubs that attract insects and simply instilling a love of wildlife and the environment in children are just as important. That's part and parcel of what we do. If someone spends a few hours here, building dens, watching butterflies, or whatever, and comes away thinking about rewilding a bit of their garden, or digging a pond for toads, or giving money to a conservation charity, then that's great. For the same reason, we encourage people to have a go at fishing while they're here. Kids especially don't always get to try fishing because the gear is expensive or they don't live near any water. But again it's something relaxing and immersive which makes people feel so much closer to the natural world. I like the fact that kids can wander around in a relaxed atmosphere, rather than worry that some fearsome park warden might appear any second. That's why we want people building dens, trying a bit of fishing – we want them to see the outdoors in a way they might have never experienced before.

Autumn, same as in spring, is a time to send the pigs into the woodland, taking advantage of what used to be called 'mast' – the seeds and nuts that virtually cascade from deciduous trees at this time of year. Pigs love nothing more than munching their way through all that natural food, with a good helping of all the beautiful grubs that lurk in the leaf matter too.

Thing is, while indoor farms have made pig meats very affordable, that's far from a pig's natural habitat. Put a pig in woodland and straight away you see how it slots seamlessly into that environment. You understand its connection with wild boar, the snuffling, rummaging, digging behaviour in its DNA, backed up massively by its incredible sensory perception. Pigs have a sense of smell so complex that they can build a picture of what's both above and below ground. Heads down, moving through the soil, they will dig and dig if they think there's something worth having underneath. Famously, they can detect truffles, the much sought-after fungi, an absolute culinary delight, which grows in the root systems of certain trees and comes to fruition in autumn. In fact, in the hope of exploiting our pigs' treasure-hunting skills, I did once plant some oak trees inoculated with the relevant fungi. In my head I had an image, twenty years down the line, of a mini-forest full of truffles. Not only that but I'd have umpteen pigs specially equipped to forage for this rare and valuable commodity. That dream was very much alive until one day I asked Tomasz, a lovely chap from Poland who'd come to work on the farm, to strim back some nettles. Maybe something got lost in trans-lation, but Tomasz didn't strim back any nettles, he strimmed back the oak trees instead – all eighty of them. Oh well, truffles are overrated anyway. At least that's what I keep telling myself! And I can console myself with the remarkable sight of honey fungus, the coral reef of the woodland, everywhere.

It's lovely to see pigs in the woods. While they can't be there for extended periods of time – they'll end up doing more damage than good – they're great for turning the soil over, unwittingly planting a lot of acorns they miss by driving them into the soil, another example of how pigs can play a hugely positive role in the woodland ecosystem. Jays are similar saviours of the oak. Drilling acorns into the autumn soil to dig up later, they have been known to forget the location of their treasure. I wonder just how many fine British oaks have sprung from their forgetfulness.

Pigs being pigs, they don't always wait for us to let them into the woods. There have been occasions where they've got through their fencing and into the trees themselves. When that happens, it's about as easy as herding cats to get them back in. Once they've started munching on those acorns, that's it, they're blind to all persuasion. All you can do is hope that at some point they'll have had their fill and head back home. On the positive side, the flavour of acorns naturally infuses the meat. In Spain, I have eaten the most incredible Pata Negra ham from Black Iberian pigs fed on acorns, and yet in the UK the link between feed and flavour is these days far less commonly known, reflecting how consumers used to be so much better connected to their food. Not only did they understand the effect of environment on flavour but the animal itself was highly valued, allowed to live longer deliberately so it could enjoy the free autumn feast before slaughter. Our rare breeds grow slower and also live longer than commercial hybrids. Those extra weeks and months mean denser muscle fibre, which in itself results in a considerably more enjoyable texture to the meat than the mass-produced alternative, which tends to crumble and fall apart more easily. There will also be a considerably higher fat content, important in the flavouring of the meat during cooking.

Back in the day, villagers would have utilised the cold weather to process and store the meat. Nothing would have gone to waste. Offal would have been enjoyed first, the best stuffed hearts, faggots fashioned from liver and kidney, and black pudding from the fresh blood. From the head and trotters, meanwhile, they'd make brawn, the feet also producing the gelatine used to set the head meat which would be eaten straight away, or possibly turned into a rough farmhouse pâté. Generationally, we're still linked to these ways. My rag-and-bone-woman grandma certainly wasn't averse to a meal of half a sheep's head with onions. I'm sure more than a few of us will have direct memories or heard stories of relatives eating tripe or keeping a pig's head in the fridge.

If it's a bull, traditionally testicles don't go to waste. I've eaten them myself, in Kenya. No messing about, they weren't on the fire for more than a minute before they were back out and in my mouth. If you can put out of your mind what you're chewing on, they're actually not at all bad.

The legs and the belly of a pig would then be cured, salted down, and turned into bacon and ham. Salting dries the meat and retards bacterial growth, which means it can be stored all through the winter. Even now in modern production facilities in Spain, the refrigeration and maturation of legs of Serrano ham still follow the yearly patterns of a traditional Spanish farmhouse kitchen. Here, however, while working with the

Next generation – every rare-breed pig that's born is important.

Producing and selling our own meat is an absolute bedrock of the farm.

Looking good in smart autumn attire – the camels, not me.

Macaques are emotional and intelligent animals.

Halloween was never more frightening than this.

Turkeys – I've spent more autumn evenings than I care to remember trying to catch escapees.

Crocodiles can seem motionless, but there's a lot more going on behind those eyes.

Eat leaves and have sex – seems to be all that tortoises do (I chose the eating leaves picture).

seasons was once second nature, today there's a tendency to pick up a packet of ham, pork or bacon in the supermarket that's been produced in a soulless and mechanised manner. Where's the connection with how that animal was raised, what it was fed, the ethos of the farmer, how it arrived neatly labelled and packaged on the shelf?

There's also the issue of food waste, a huge problem in our modern world. According to the Prince's Foundation's Food for the Future programme, some 1.3 billion tonnes of food are wasted every year around the globe. That's around a third of all food produced. Prince Charles's initiative, which I've been happy to join, aims to demonstrate to the next generation how the food production system works, and equip them with the skills and knowledge to reduce waste. Remarkably, tackling this issue could reduce global greenhouse gas emissions by up to 10 per cent – food left to rot in landfill is a major source of methane. In fact, food waste is even more dangerous for the global climate than plastic.

We sell rare-breed meat and sausages from our pigs in our farm shop and restaurant. Its backstory is there for all to see, on menus and information boards. Pleasing to report that there's something about reconnecting with traditional food and the way in which it's sourced that appeals to a rapidly increasing number of people who want to remove themselves from the conveyor belt of blandness that so much food has become. I'll never forget – believe me, I've tried – the grease-laden slop served up to me under the laughable title of 'full

LIVER AND BACON

Serves 4

Bacon – properly cured bacon – is a national delight. Liver, on the other hand, we've fallen out of love with. A real shame, because not only is it super-delicious and full of nutrients, but to dismiss offal, such an incredibly versatile part of an animal, is incredibly wasteful.

Some people are put off liver after being served it overcooked and rubbery at school, but actually, done properly, it's fantastic. Just think of it as pâté that no one's messed around with!

I challenge anyone to cook this dish and not do it again and again.

Ingredients

750g lamb's liver
plain flour seasoned with sea salt and freshly ground
    black pepper
a little oil, for frying
12 rashers of dry-cured streaky bacon

Method

Give the liver a quick rinse in cold water and pat dry with kitchen paper.

Slice the liver about 1cm thick. Remove any membranes or thick tubes.

Lightly coat the liver slices in seasoned flour, shaking off any excess. Set aside. Then heat the oil in a non-stick frying pan and fry the bacon on a medium heat for 15 to 20 minutes until crisp and golden.

Remove the bacon from the pan and keep warm, then add the liver to the remaining bacon fat in the pan. Fry for a couple of minutes on each side – liver cooks very quickly so make sure you don't overdo it. Serve the liver and bacon with mashed potato and onion gravy.

English breakfast' at a motorway service station. Losing my rag really isn't my thing but a head full of jetlag combined with a price tag of £9.80 meant it was more than I could take. The manager gave me some corporate spiel but the contempt for good food, customers and the breakfast itself still leaves a bad taste in my mouth. The full English – fried egg, black pudding, bacon, sausage, beans, mushrooms, fried tomatoes, with a big pot of tea – is a classic and, to be honest, not that easy to get wrong. I cannot understand how any organisation would wish to associate itself with such awful produce, let alone serve it at such a hyper-inflated price.

I don't think it's good enough for someone to put a tray of meat in a trolley, or a burger under the grill, and say they like it but don't want to know about where it came from. It doesn't matter what we eat, we need to be honest with ourselves. If I'm eating a vegetarian meal, I'm aware that livestock have enriched the soil in which those vegetables have grown with manure, same as I understand pests such as caterpillars and pigeons will have been controlled. Eat even a salad and chances are something will have lost its life through displacement, habitat change, or pest control. It's practically impossible for anyone to say no animal was harmed in the making of their food.

It's too easy to wear a blindfold and ignore food's provenance. Over time, it feeds an ongoing lack of respect for and understanding of farming and food production and a complete miscomprehension of what we're putting in our mouths. Enjoying the natural harvest is something else, unwittingly, in just a few short generations, we have become disconnected from. I was talking to a woman recently who, while out on a walk, had picked an apple off a tree and bitten into it. Her partner couldn't believe what he was seeing.

'What are you doing?'

'I'm eating an apple.'

'What, straight off the tree?'

Occasionally, I see a child pick a strawberry and eat it, and then hear their parent ask, 'Is that all right?' It hadn't been ticked off some checklist as being OK to be sold; it hadn't been washed. It's because we've become so accustomed to fresh food being packaged up in supermarkets.

I've seen the same thing happen when I've offered someone a runner bean fresh off the plant – 'I can eat it like that? It hasn't been cooked!' It shows how big the disconnect has become between us and our food, us and the natural world. And that has all happened within two generations.

Because we've handed so much of our food intake over to someone else to take care of, it's become the norm to think an apple that hasn't been processed and bagged up is in some way not fit to consume. I found these stories particularly sad because to be in touch with nature is a wonderful thing. It's lovely to be able to see a cloud formation and understand what it means; to see some cuckoo spit and know its purpose (to protect the tiny baby leafhoppers that lurk within); to read the environment and understand its little cues. It makes you feel so much more connected to nature rather than separate from it.

We try to practise what we believe. The meat we eat comes from the farm and as much of our fruit and veg as possible. That's massive to us and of course we're in a perfect position to show our children the history of the food on their plates. But on a wider scale, I hope the idea that food is something that comes in a packet from a shop can be reversed. All children should know where their food comes from. There needs to be a practicality to that. If a child sees food as something that comes from a shop or out of the freezer, then the result is a lifelong disconnect that is good neither for them nor the environment.

We humans overcomplicate just as much as we oversimplify. How much time and money is spent chasing the latest food trend, the one that's going to leave us healthier than ever before? Is that really the answer? Previous generations ate food from the greengrocer and the butcher – just as we are told to do now, and it was produced in a straightforward way, from the land, better for them and the animal.

The breeding and use of pigs as a quick and cheap food source goes against the entire natural spirit of the animal. Pigs are meant to be active, to enjoy the open air; they are intelligent sentient animals able to feel happiness, sadness, and deliver affection. When I see our Middle White piglets running around in the autumn woodland, I'm transported back through the decades to when this sort of thing must have happened all the time and I'm proud to be part of a re-emergence of traditional pig-rearing methods.

While autumn might be a little late for showing my cooking skills on a barbecue, I can still satisfy my primeval urges with a bonfire. Akin to the first shaft of spring sunshine, a bonfire is one of the great sensory indicators that autumn is upon us. There's something really lovely about raking through the garden waste, making a stack and then watching as the smoke lingers and mixes into the cold air. It always reminds me of jacket potatoes cooked in foil – opening them up and letting a big blob of butter melt inside. One of the great tastes of the autumn.

Most of the waste from the garden, farm and wildlife park we like to compost. That way, just like the woodland does so naturally, we are recycling organic matter back into something usable. Just as the leaves dropping from the trees feed the bluebells in the spring, the vegetation we pull up or cut back becomes the nutrients supporting the next cycle of growth. Compost is an amazing resource. It tends to be the more woody stuff that can't be composted which goes on the bonfire.

Another great signifier of the autumn is my trip to the orchard to pick the last of the pears. In the early morning, walking through the grass, cobwebs sinking under the weight of the dew, I'll be reminded of fruit-picking as a kid, particularly blackberries, half of which I'd have eaten before I got home. Maybe I think too much but I can't help comparing myself to the butterflies, bees and small mammals making the most of the last berries, flowers and nectars. In the wild, nothing is wasted – everything that can be is devoured.

Like childhood summer holidays forever remembered as sunny, in my head, autumn pear-picking always happens in sunshine dappled by the branches of the fruit trees. Last autumn, I taught our youngest daughter, Bo-Lila, how to pick a pear. The bottom of the fruit is called the eye.

'Turn that to the sky,' I told her, 'and if it's ripe the pear should just snap off.' She did just that, and we ended up filling bag after bag, keeping a few of this great autumnal glut for ourselves, even having a chomp of one or two of the juicier-looking ones. I couldn't help thinking of my own childhood and family strawberry-picking at a farm near Bishop's Stortford. I'm sure I won't be the only person to recall such trips as a chance to stuff your face with fruit before taking a measly half punnet back to the entrance to be weighed.

Bo-Lila and I dropped the remaining fruit at the restaurant, where the chefs got busy peeling and chopping to make delicious crumbles and pies. A pear and raisin crumble with custard is the food of the gods.

We'll go through the same routine with the apples, anything that doesn't go home or to the restaurant providing a tasty autumn treat for the animals. We have so many that even the donkeys get fed up after a while. Goats, on the other hand, are a different matter entirely. If they spot an apple tree it can be your worst nightmare. They have a reputation for eating everything for a reason. It's not just the apples they'll consume, they'll strip a tree of leaves, climb it and pull the bark from the trunk. They've evolved to make the most of the food resources in any given area. A diet of pretty much anything gives them a better chance of survival. They're not the only ones to look out for. Muntjac deer, a non-indigenous species, originally from Asia, are present in the area. Given the chance, they'll also make for the apple trees and eat the bark, especially late in the year when other food sources become scarce.

We have waste food, fruit and veg, come to us from supermarkets, always useful for feed – tapirs have very little interest in whether an apple is bruised or wonky. On one notable occasion, we had 5 tons of mushed-up pulp, pips and peelings from a jam manufacturer – the pigs were in absolute heaven, managing to get through their share of the mound before it started to ferment. Pigs can be unruly enough without being transformed into four-legged football hooligans. It was good to present the pigs with such a bounteous treat. In the early days, they had played a vital role in helping us transform an overgrown wreck into something better resembling a farm. It never hurts to say thank you.

Sometimes it's the less expected animals that seize an unexpected opportunity to gorge. On one occasion, we had an entire container of potatoes from Felixstowe, rendered useless for human consumption because they'd split open. We piled them up and erected an electric fence to keep any nosy animals away. Naturally, the fence went down one day and one of our oldest Red Polls, Missy, a good traditional all-purpose cow, was straight in there. Next time I saw her, I couldn't believe how well she looked. Knocking on a bit, she tended to look her age. Now, though, she looked amazing, like she'd really found a bit of condition from somewhere. But then the closer I got, the more I thought she looked a little odd – not so much weightier as distended. I could also see she was salivating a little. We got the vet in and with a little prodding he discovered the issue – a potato had become lodged. A cow is basically a massive gas tank. It eats a huge amount of vegetation and as all the bacteria get to work the by-product is an awful lot of gas. There are two ways it can come out. If one is blocked then there's a problem, as exhibited by Missy, who was getting bigger and bigger by the minute. The vet had a piece of pipe and, after asking Missy to open wide, he managed to poke the potato down into the stomach. Obviously, this entailed him having to stand right by Missy's mouth. The second that potato was dislodged an almighty rush of wind erupted from inside the cow. I swear it parted the vet's hair on the way out. After this vast bovine burp, Missy just wandered off like nothing had happened. In my head, she'd gone from a calm and docile cow to something that could flatten entire mountains.

During the pandemic, when the restaurant was shut, we were especially overwhelmed with apples and so we collected them

WINDFALL AND BLACKBERRY CRUMBLE

Serves 4–6

Windfall apples, bushes abundant with blackberries . . . It would be rude not to take advantage of this luscious harvest and rustle up that great British creation: the crumble – not so much a pudding as a warm hug.

What I really love about crumble is not just its incredible texture and taste, but its sheer versatility, making it the perfect way to turn fruit that might otherwise be wasted into the most wonderful treat, precisely the reason such puddings were invented years ago. Pears, plums, apples – I'll use anything I come across.

After I've been out collecting, I make sure that I freeze wild and windfall fruits as soon as possible; that way, when I hear the call of the crumble, I've got bags ready to go.

Similarly, the crumble itself can include anything from old biscuits to (if you fancy a Christmas crumble) leftover mince pies and chocolate coins. Give it the crunch you want!

Ingredients

225g plain flour

115g butter, cut into cubes

85g demerara sugar

600g chopped (peeled and cored) apples (cooking apples work best as they hold their shape, but this is a great way to use up all unwanted apples)

2–3 tbsp caster sugar

225g blackberries

Method

Preheat your oven to 200°C/fan 180°C/gas 6.

To a large bowl, add the flour and butter. Rub the mixture together with your fingertips until it resembles breadcrumbs – use a food processor if you want. Mix in two-thirds of the demerara sugar and set aside.

Now put the peeled, cored and chopped apples into a shallow ovenproof dish. Sprinkle with the caster sugar and cover with foil. Bake in the oven for about 15 minutes until the apple starts to soften, then remove from the oven, stir, and add in the blackberries. Ideally, you want between 4 and 5cm of fruit. Sometimes, just to give it a little bit more liquid, I add a splash of hot water to a nearly empty jam jar (any flavour will do) and add it to the mix.

Cover the apple and blackberry mix with the crumble, making sure you have an even layer, then sprinkle the rest of the demerara sugar over the top. Bake for another 30 minutes. By now the fruit should be starting to bubble up and erupt through the surface.

Serve with lashings of custard or your favourite ice cream.

up and gave them to the pigs, a reflection of the old farming system where you'd run the pigs through the orchard to pick up all the windfalls. I cherish the language surrounding nature, and 'windfall' is a great example. We use the word in everyday life to describe an unexpected bounty, and in nature that's exactly what it is. That fermenting fruit is heaven not just to pigs but to butterflies, wasps, all kinds of insect life, a melting pot of sugars. Of course, it's also great for ammo in a fight with your mates.

We have seven different breeds of pig on the farm. To see them munching on the apples amidst the last of the autumn sun during lockdown is a stunning memory from a difficult time. It was also a reminder of these great animals' heritage. The Gloucestershire Old Spot was once known as the orchard pig, its signature pattern said to be bruises from where apples had fallen on its back.

The absence of restaurant diners meant the end-of-season goodness from the garden went the animals' way too. I can now exclusively reveal that turkeys love runner beans, cows enjoy Swiss chard, and capybaras are massively up for a pumpkin, grown both for the restaurant (who doesn't love an autumnal pumpkin soup or spiced roasted pumpkin?) and our annual Halloween pumpkin hunt, when children are also invited to encounter frightening ferrets, mischievous macaques and Pebbles the scary sheep – it's true, sheep really can be scary! I try to use every bit of the pumpkin, drying and roasting the seeds for salads or even holding on to them to plant the following year.

At this time of year, trees and hedgerows also act as natural fodder for a lot of the exotic animals. While sheep are grazers, heads down hoovering up the grass, goats are browsers, picking nice little bits off trees, bushes and hedges, here, there and everywhere. We cut parts of willow trees for the camels. They absolutely love eating the leaves and stripping off the bark. In the wild, that's exactly what they'd be doing. In captivity, we bring it to them, their own wildlife park delivery service, a great way of utilising our woodland as another food resource.

Feeding the animals is, naturally, top of a long list of daily tasks. Trouble is, in autumn, the days get shorter and shorter. I have nothing but admiration and respect for everyone who keeps the farm and wildlife park running smoothly, because there are times when it seems like an impossibility. Just when you think you're on top of everything, something else pops up to consider. For instance, our cattle have to be routinely tested for TB, a huge job which sucks up staff needed elsewhere. TB testing is a time of great nervousness. A positive animal must be taken for slaughter, the absolute last thing any farmer wants, let alone one who keeps rare breeds. You can be as confident as you like that none of your cattle are infected, but until the vet delivers that clean bill of health, there is always a little bit of worry. Only when they say the magic words do you suddenly realise just how much tension you've been carrying while waiting for the results.

In autumn 2020, we had something else with which to contend – something that had the power to pull the plug on everything we do here. Every three years, we must undergo an inspection in order to keep our zoo licence, something that's mandatory as soon as you start keeping non-domestic animals. The run-up to an inspection can feel quite fraught. Ultimately, an inspection team, consisting of vets and those with experience of wildlife park management, can take our licence away. A wildlife park has to be run to the absolute highest standards. Your buttons have to be polished, your t's crossed and your i's dotted – and that's exactly how it should be.

Inspectors have the authority to look into every nook and cranny – bins included – and so we have to be up to speed in every conceivable way. But the actual inspection is just the tip of the iceberg. Underneath is the paperwork – conservation policy, education policy, escape policy. There are strict regulations surrounding keeping wild animals and any infringement regarding medical records, keepers' diaries, enclosures and a million and one other things means a licence can be withheld and the entire operation shut down.

While we're always 100 per cent confident that we run the park to the highest standards, an inspection does feel like the exam of all exams. Driving test, GCSEs, A-levels, none of them come close to having several, albeit very polite and well-meaning, individuals holding your future in their hands. The inspection of autumn 2020 was particularly nerve-racking. The park was four times bigger than when our licence was last issued and preparation work would have to be completed alongside the day-to-day jobs that already seemed to consume every minute, not forgetting we were also filming a TV series for Channel 4. It says everything about the commitment of the team we have here that we passed the inspection and received a six-year licence, the inspectors commending our 'enthusiasm, dedication, and determination to improve and progress'.

The inspection is a serious business but I couldn't help being somewhat amused by a comment in the 'public safety' section of the report regarding an 'overconfident' rhea being able to interact with visitors. Rheas are large, flightless birds with grey-brown plumage, long legs and long necks, similar to an ostrich. They have an amazing quirk in that during the courtship display, the male produces a deep and thunderous roar. You're looking round wondering who ordered a lion when you realise it was actually a bird. This particular rhea, Geoff, is, admittedly, a show-off. He's a male and just wants to display all the time, puffing up his feathers, and coming right up to the fence so everyone can see what he's got. To be fair, this kind of behaviour can happen in rescue animals that have been cooped up in backyards or gardens, often on their own. Such isolation means they identify with people much more than they would in the wild, possibly even seeing us as rivals. In the face of any rival, an animal's immediate instinct is not to look weak but to be 'overconfident' as the inspectors saw it, as if he was a touch cocky, a bit of a wide boy from Essex.

Could this be the start of an inspection trait? 'This wildlife park has an exceptionally arrogant camel.' 'Their crocodile really does need to wind its neck in.' It made me think of an Ofsted inspection, when schools have been known to send troublesome kids on a trip so they don't hijack the report.

Unfortunately, I don't think you can send over-confident rheas to Alton Towers for the day, joined on the coach by a mischievous giant African land snail and a recalcitrant Mandarin duck.

With our own future secure, the next job was to make sure our animals were as comfortable as possible for the approaching winter. That meant, amongst other things, giving an emergency haircut to Mystic, one of our guanacos, the wild relative of the llama. Mystic's growing of a much-needed new winter coat was being thwarted by the fact he was still wearing a shaggy old one. Guanacos, with their camel-like eyelashes, which make them look like they're about to take a selfie, are outwardly gentle creatures. However, they will spit when threatened – and when I say spit I actually mean projectile vomit. They can hit a target from six feet and, disgustingly for anyone who gets in the way, have a very good aim. With this in mind, no way was I going to go bumbling in there with a big buzzing pair of electric clippers. Hand clippers, used with speed and dexterity, were definitely the order of the day. Mystic definitely wasn't happy, but I'm pleased to say he saved a volley of projectile vomit for another day.

While some animals are being readied for colder days ahead, others are still moving in. There's always an element of excitement when a trailer pulls up, but this particular vehicle caused a real buzz, containing as it did two black-and-white ruffed lemurs and four ring-tailed lemurs. Immediately, they were a big hit with our girls – not only are lemurs very friendly but they look like fluffy teddy bears. They're perfect for a wildlife park because visitors can walk among and interact with them. In the wild, ruffed lemurs are remarkable for being the largest pollinators in the world. Pollen sticks to their ruff as they feed, which is then spread from tree to tree. Both our species are endangered, the latter critically, and are native to Madagascar, where wildlife habitat is being destroyed at an alarming rate, with 90 per cent of native forest cover already lost.

The arrival of the lemurs was part of a bigger plan to create a Madagascan exhibit, splitting the existing butterfly house in two so the lemurs would have an indoor enclosure leading on to a bigger area outside. Initially, there was discussion about where this rejig would leave our movie-star goldfish that came to us after being featured as extras in *Spider-Man: Far From Home* – would they be able to remain in their butterfly-house home? I mean, you can't go moving Hollywood stars around. Before you know it, you've got their agent on the phone. In the end, they stayed. Although maybe it wouldn't have mattered. They have, after all, got very short memories.

The lemurs are all rescue animals and, having lived in a small cage, it's great to see them revel in having space in which to exercise and explore. To add to the Madagascan feel, we are also introducing birds native to the area. As always, we want where our animals live to feel and look as naturalistic as possible. We can then add information boards so visitors are aware of their plight. I find lemurs particularly fascinating because they tell an alternative story about the evolution of ourselves.

Lemurs advanced in the total absence of monkeys and so are essentially a parallel alternative. It makes me wonder, had we too evolved in a world with no monkeys, what would be the 'zero monkey' version of us? And would I, if I was closely related to a lemur, still have opened a wildlife park?!

For us, the end of the season, when it's quieter, becomes a lovely reflective time. While there are still plenty of jobs to crack on with, like getting the garden in shape, and endless pruning, there's a period of calm where we feel as if we almost have the wildlife park to ourselves. Making plans has become part of that.

An Indian rhino is my big dream. Rhino are commonly thought of as being African, but actually most species are from Asia. I was lucky enough to see an Indian rhino on safari in Nepal. While I'd been impressed by what I'd seen on documentaries, when we encountered a group (known, a little bit disturbingly, as a 'crash') in the hazy evening light, I could barely comprehend what was in front of my eyes. These vast and majestic animals made me wonder if I'd somehow travelled back to prehistory. They were like a four-legged tank, clad in armour. I had a mental picture of them getting home at the end of a hard day and sighing with relief as they took it all off. My vision is to create a large natural enclosure and bring an Asian rhino here so people can see for themselves just how spectacular an animal they are. With numbers in the wild down to the low thousands, it's vital that we do as much as humanly possible to educate people about their fight for survival.

Again, if we were to bring in a rhino, we would want to place it by geographical region, same as if we brought in another deeply impressive animal, the brown bear, rivalled in size only by its Arctic relative, the polar bear. A brown bear is particularly exciting as we could incorporate woodland in its enclosure and maybe add bison, European elk and potentially our reindeer.

Penguins too are on the horizon. For a while now, I've been considering the viability of replacing the side of a shipping container with a viewing panel, essentially turning it into a giant water tank. Shipping containers were the answer with the macaques, so why shouldn't they be again? They're an amazing reusable material.

Talking of water, I would love to have an aquarium. I've kept fish all my life with varying degrees of success (see earlier!) and at present we have some koi carp at the wildlife park but they're not exactly a main exhibit. When I've asked a family getting out of their car what they're hoping to see, I've yet to hear anyone answer, 'It's obvious, isn't it, Jimmy? The koi carp, of course!' The problem with an aquarium is having the indoor facilities in which to house it. Building from scratch, with all the expense that entails, would be the only answer. Maybe one day, though. There is, after all, something so lovely about walking into a dark aquarium and seeing a big fish tank glowing, full of the most incredible sealife.

Otters too, such sleek, beautiful and incredibly watchable creatures, are on the list, dependent on finding the right home for them – I'm thinking maybe not too near the aquarium.

●

A rhino, or any other significant addition, isn't just going to arrive overnight. We see what's happening here as a very long journey and in some ways, after twenty years we're only just beginning. We think about what's working and what isn't and how we can build success and change into the long term. Last autumn, for instance, we talked about the conservation and rescue element of what we do. While a lot of the animals are here because we've rescued them from facilities that can't look after them, or they've been privately kept in an inappropriate manner, we want also to aim towards rehabilitation and release. Already we take in injured hedgehogs, restore them to good health, and release them into our woods, but we want to do something similar with our exotics as well, potentially with Asian fishing cats, hopefully for release back into their native Sri Lanka home.

It would be easy to sit back and be content with what we've achieved, but I don't think I'm ever going to be that kind of person. I'll always want to move forward and develop what we're doing. I've tried standing still and it doesn't work! Michaela and I had a chat where we agreed no more animals for a couple of years, but then the lemurs needed rescuing and we were right back in the thick of it. Someone else contacted me about giving some flightless macaws a home. Once an idea like that is in my mind, I find it hard to turn my back on it. Admittedly, some ideas are a little more far-fetched than others. Yesterday, I had a sudden thought – how difficult would it be to net the entire Outback Safari area so we could have the kookaburras flying around in the same space as the emus and add another Aussie, the lorikeet, a bright and vividly coloured member of the parrot family, to the mix? In Jimmy World, there's always room for one more.

Truth is, even with the wildlife park expansion of recent years, we still do have lots of space to broaden further. I honestly think we've barely even tapped into what we've got yet. Many collections are restricted in size by virtue of the fact they are located in city zoos. We're lucky to have space. It truly is our most precious commodity, because we can think not only about bringing in certain animals but how we can give them the most dynamic environment, be it one large enclosure or a series of small interlocking paddocks.

Our farming operation has also changed. The food we produce these days, turkeys aside, stays on the farm, sold through either the shop or the restaurant, which is lovely because the food miles are practically zero. It also means we can ask ourselves how much food we actually need to produce and if there are parts of the farm that can be turned over to the wildlife park.

Chasing potential is a lovely thing because it drives the place forward, but it can be a curse as well if you overreach, get up one day, and all the money is gone. It never fails to amaze me how much a step forward can cost. Recently, we laid an electrical cable from the main part of the farm down to the butterfly house. The bill was £20,000. I'm fortunate in that my mindset has never radically switched from my original objective, to put into practice what I believe about conservation, preservation, protection and welfare. Later that extended to giving my family a lovely way of life, to letting my children know about the natural world and where their food comes from.

Thinking sustainably is the key to longevity. I take a lesson from our hibernating animals. They're careful to build their reserves to a point where they know they can survive. Nowadays, a little differently to that naive young bloke who turned up two decades ago, I'm careful to make sure we can afford to invest before we commit.

Some ideas are more out-there than others. We have staged festivals here in the past, and may yet do so again, using it not just as an excuse to listen to some great music but as a tool for helping the environment. I'd love it if, for every person who came along, we planted a tree – I think people these days are conscious of not just wanting to enjoy themselves but knowing they've done their bit. It would be quite amazing to come away from a festival knowing you've helped to plant a beautiful oak woodland. I believe in the idea of people being able to let off steam – it's another great countryside tradition. Years ago, people didn't have holidays, maybe not even a day off, but they did have a harvest festival or a village fete where for a few hours they could forget the hard toil of daily life and let their hair down a little.

However realistic or preposterous, I love that autumn element of making plans and gathering thoughts. It's as if the lowering of the sun, the changing of the light, automatically flicks a switch that makes us think of the future, ready for the big surge again in the spring. Before any of that can happen, though, first must come Christmas. 'Why mention that in the autumn chapter?' I hear you ask. Well, because early autumn is when 8,000 turkeys come to stay, enjoying top-quality bed and board for a couple of months before the inevitable and their transfer to the supermarket shelf. Our polytunnel becomes the avian version of one of those old hairdressing salons I used to pass as a kid. The noise of thousands of turkeys, amplified by the acoustics, is ridiculous – a load of old ladies nattering away as they wait to have a blue rinse. They are also immensely inquisitive. I'm convinced if I fell asleep in there, I'd be found a few hours later pecked to the bone.

Polytunnels are great for providing a settled environment – space, food and comfort – for turkeys twenty-four hours a day. Turkeys are sensitive to even the smallest change in their surroundings, be it temperature, light or diet, and a negative reaction can prompt them to peck and injure one another. Once they're used to their home, we open up the sides so they can range out in the pasture. At night, we have the fun task of chasing them all back in again. We're not here to play Santa Claus for foxes, although you might have thought the opposite when we had the nightmare of a mass escape of a thousand birds onto neighbouring land. Rounding up a thousand turkeys, by the way, is a job I wouldn't wish on my worst enemy. I wouldn't have minded if I'd never seen another turkey in my life – although I soon changed my mind on Christmas Day!

The arrival of the turkeys signifies kick-off for the run-up to the festivities. The demand for free-range birds has gone up in recent years. Christmas dinner is, after all, the most important meal of the year – you can buy all the presents in the world, but sitting round the table with your family is priceless – and more and more people are seeing how much more flavour is added to the meat by allowing a bird to roam as nature intended and grow slowly, as compared to the faster-developing hybrids that make up the vast majority of mass-produced commercial turkeys. Of course, there's an impact on price, but think about it in terms of cost per portion and you're looking at about three quid – much less than a pint or a box of chocolates – and that's without the delicious pies, curries and sandwiches that come from all the leftovers. In fact, we actually run Christmas turkey carving classes to show people how to get every last tasty morsel from their bird. Many shoppers are tempted to buy turkey crowns, essentially the bird minus legs and wings, seeing it as a cheaper alternative. But to leave just the crown, the entire turkey still has to be butchered, a cost which is passed on to the customer. Go down that route and essentially you're paying for the whole bird and receiving only part of it. That's not just a Christmas tip. Buy a whole chicken or fish at any time of year and you reap the benefits. As well as the extra cuts, you have the bones and carcass for soup or stock which can be used to make risotto.

As well as the temporary intake, we do have rare-breed turkeys on site all year round, including Nick, a member of the Slate breed, named after its ash-blue colouration; Dom the Bourbon Red, a North American breed with striking russet brown and white feathers; and Gerard the Narragansett, named after the bay in Rhode Island where the variety was developed. All are on the Rare Breeds Survival Trust watchlist, with urgent action needed to prevent extinction. My plan is to create three separate flocks combining natural mating and artificial insemination, which ups fertility from 70 to 95 per cent, but does tend to produce a flood of complaints to Ofcom when scenes of me stimulating a turkey behind the parson's nose (the protuberance at the posterior end of the bird) and then using an old-fashioned suction method to extract the sperm is shown first on *Autumn at Jimmy's Farm* and then repeated on *Gogglebox*.

By now, with the dark nights drawing in, a veil of darkness descending on the park, breath visible against the car lights of the visitors heading home and the invitingly cosy glow of the restaurant, it really feels as though winter is almost upon us. This brief period is one I have always loved. It holds in its hands an expectation, an excitement, even a delightful trepidation about what winter might bring. It reminds me of coming out of school and tending to my animals while the house warmed up, of evenings in my bedroom endlessly reading about animals or watching my menagerie with the curtains shut.

It's odd in a way that the process of saying goodbye to autumn should feel so welcoming, so embracing – a beautiful gateway to the hardest season of them all.

Winter

A harsh season with a stark beauty all of its own.
A time when the natural world reveals astonishing acts
of survival set against the most mesmerising of visual feasts.
A time to toast the year just passed and welcome
with open arms the one to come.

Running a farm and wildlife park can never be a half-hearted pursuit. You're either in or you're out. There's very few things worse than trying to straw up pigs on a freezing-cold night in November when you've got flu, it's lashing down, it's cold, there's no money coming in, and – as in the early days – all you've got to go back to is a caravan with three dogs and a pig. YOU REALLY HAVE GOT TO ENJOY WHAT YOU DO! Because sometimes you take a look around and it feels like you've voluntarily sent yourself back to the Dark Ages.

I clearly remember the first snowfall we experienced here – because it came on the exact same day we opened the farm shop. We had dressed it up to really look the business, with a down-to-earth countryside feel, ready for the big push – and then down came the white stuff. I couldn't believe it. The road in was inaccessible, the water pipes froze – calamity after calamity.

But Fortune was on our side and by the following morning the worst of the snow had gone. We had a really busy day and celebrated down at the pub. On returning about 11 p.m.,

I glanced through the shop window only to be confronted with a strange *Twilight Zone*-esque mist. The snow might have gone but it had been replaced by really heavy rain. The entire shop had flooded, three inches of water everywhere. While bed beckoned after a long day, we had no choice but to stick Bon Jovi on the CD player – 'Livin' on a Prayer' could have been our motto at that time – get in there, clean up and get everything prepped and ready for the morning when we opened up at – Oh no, really? – 8 a.m. Thankfully, I had a digger and so could dig a ditch to stop more water coming through, but it was a miserable task when all anyone really wanted was to be curled up in bed, one of those occasions when every fibre of your body wants to scream and shout, but underneath you know that, actually, what's the point? It won't change anything. You just have to get on with it. Tomorrow's another day.

Shop-opening day is the only occasion I have ever truly cursed the snow. Any other time, I absolutely love it, even if it rarely heralds good news for the farmer, with its propensity to incapacitate pipes and equipment. Frost is the same. Often seen as a negative, the first sign of winter, when in fact in nature's never-ending cycle it actually kills off a lot of bacteria and therefore prevents disease. On a more aesthetic level, frost is also so lovely on the fields. I open the curtains in the morning and it's as if I'm looking at a scene that's been placed there, quite deliberately, brand new out of a packet. Frost, like snow, hides the rough patches. It has a cleansing quality, taking away the harshness and making everywhere pristine.

On a frosty morning, I like walking around the wildlife park before it has opened up. It's such a lovely experience. Everything feels so fresh early on, as if it's just been unwrapped – unsullied, unspoiled. I'll be blown away by the intricacy of ice crystals glued to the twigs and reeds around the pond, the thinnest film of ice on the water. The legend of Jack Frost makes so much sense because it truly is as if an invisible hand has come in the night and turned our little corner of the world into a wonderland. I look up and see that big open sky that frost always brings, soon filled with the corvids – the crows, rooks, jackdaws and beautiful jays that roost up in the woods. Magpies are among them too, reminding me, as ever, of 'one for sorrow, two for joy'. I stop before 'three for a girl' – I've got four already! Even after all these years, Michaela still insists on saluting the magpies, believing doing so is a way to ward off bad luck. Seagulls, meanwhile, always seem to see an empty sky as something to fill, making the short trip over from the nearby coast.

There's only so much open-mouthed gawping you can do at a winter landscape. The work doesn't stop. Some people see snow-capped hills, I see snow-capped humps on my Bactrians. Glad to report, these camels are hardy creatures. While it's easy to think a lot of wildlife park animals would, in their natural habitats, be living in heat, truth is many live amidst great extremes of temperature. Bactrian camels, for example, as with a lot of desert animals, tolerate an 80-degree range, from plus 40°C (104°F) in the extreme heat of day to a hoof-curling minus

40°C (–40°F) at night. Never mind all the incredible all-terrain vehicles you see on *Top Gear* – the Bactrian blows each and every one of them away. Not only can it deal with ridiculous cold and heat, but it can weather sandstorms, strong winds, rocky mountains, stony plains, dunes, drought and lack of fuel as a matter of absolute routine. Look at the Bactrian camel and you're seeing the SAS of the animal world. Forget the electric revolution, we should all have one of these sitting in the drive, with the added bonus of a little something for the flower beds.

We never get down to minus 40 in these parts – it just feels that way when you're leaning at 45 degrees into a headwind in January – but we have had some desperately cold snaps down the years. The first thing to happen in that situation is the water freezes. We can't ever have a situation where the animals don't have access to water, and so suddenly we're having to physically ferry the stuff around the park – which is when you realise just how much a cow can drink. Fill up a trough and it's gone in seconds. Fortunately, while I think I'd look quite fetching in the outfit, I'm not replenishing water stations milkmaid-style with two buckets and a yolk. I have my old Land Rover to help me out, an essential piece of winter kit, although even with the classic farmers' four-wheel drive, there are times I'll get stuck. On one occasion, I was taking water down to the cattle when a blizzard descended. In the absence of International Rescue, I had to call on the kids, not too happy to find themselves standing in the freezing cold holding down a fence so I could make my escape. Water-carrying might only be part of the answer, of course. If the pipes have burst, you've then got to find all the leaks. Often, the actual snow isn't the real issue – it's the thaw. A hard frost is preferable to snow because the ground remains hard. When snow melts, you're left with a quagmire. For every lovely crisp winter day, so invigorating, when you can't wait to get out there, there's those others where you can't avoid the truth of the situation – everywhere is just damp, wet and cold. Mud, mud, and more mud.

It's a side to the park that people don't see, how unsympathetic the weather can be for those who work here. I know from my own experience how tempting it is to stand in the office with a hot cup of coffee, as opposed to filling a bucket with ice-cold water and traipsing a quarter of a mile to a pigsty. Winter's a testing time for the staff. It really shows their commitment. Nothing stops for the weather, simply because it can't. You've got to get out there. The animals need sustenance, not only to keep their bellies full but so they can generate the heat that carries them through the cold. And each job has to be done with just as much care and attention as it would be on the warmest summer's day. Take a shortcut to try to save a few minutes and you can guarantee it will come back and bite you in a way that would make a sharp nip from a crocodile feel positively delightful.

For my part, I try to make sure I'm equipped for all eventualities. Wintertime is when the waterproof jacket comes out, coupled with waterproof trousers worn over – I repeat, *over* – some welly boots. Tucked-in waterproof trousers are the surest sign of someone who doesn't spend a lot of time on a farm. They are missing out on a great time- (and soap-)saving trick – get back in, unzip, and you're clean. It's as close as I get to being James Bond. My version of that moment in *Goldfinger* when 007 climbs out of the water, unzips his diving suit and reveals a perfectly pressed tuxedo underneath. Ranger Tom, meanwhile, will be out there in all weather in shorts. Wrong on multiple levels. I simply can't understand what he's thinking! Having said that, there are times when, no matter what your choice of garb, you have no choice but to get filthy and really not care about it. After all, James Bond did a lot of things but he never had to dig out a bog to re-establish a swimming pool for a tapir. He also didn't spend several hours in an icy wind, finally getting home with purple cheeks that looked like they'd been on the wrong end of a fight with an industrial sander. Even Bond, though, would struggle to look good against the Italian (always the Italians) tomato producer I met recently. He had massive sideburns, cool trousers, braces, crisp white shirt. I had a knocked-out old pair of jeans and a T-shirt four times too big. Italian farmers look like they've just come back from the tailor.

A slice of Italy was certainly lacking in our first winter here – unless there's a little-known part of the country that features second-hand caravans marooned in a sea of mud. We did at least have little electric heaters, which actually made van life quite cosy, while also covering the windows in condensation.

At that point, Michaela was still commuting to London. I'd drive her to the station, bomb back to the farm, head up to the hill overlooking the tracks and wave as the train went past. She could see me but on a train whizzing past at 70 mph I had precisely zero chance of seeing her.

Michaela's days started in a manner pretty far removed from those of her work colleagues. She'd fill up the shower tray with hot water the night before, add a dash more in the morning and use it like a mini bath. That worked OK unless the plug was slightly off centre. Waking up to an empty shower tray suddenly reduced our one-star accommodation even further. Lucky for us we were young and naive enough to see the whole thing as oddly – very oddly – romantic.

Amazing, though, how much you get done living such a basic existence. There were so few distractions. Life at that point was just about feeding and watering the animals or getting on with some repair and renovation work if the weather allowed – and I loved it. There was something very visceral about it. Forget dinner parties, gym membership, dental appointments, it was all about filling buckets of water and chopping trees down, the most basic fundamentals. I should really have been around 400 years ago and lived on the Isle of Ely.

If worst came to worst, we'd hole up, have a cup of tea and watch a film. And most of the time in our winter refuge we were warm. The key thing was never to open the door – which becomes a little problematic after a while. Those mornings running Michaela into Ipswich could be a shock to the system, going from under-the-duvet snugness to the pitch-black of a winter morning, generally to an accompanying bang as a winter gale slammed the door against the side of the caravan before careering through the interior like an icy vandal.

She made that unenviable commute for almost a year before making the momentous decision to turn her back on the TV industry and throw everything in with me here. By that point, the business was rapidly expanding. The shop was open and Michaela made it fly by sourcing local products and sorting out its finances, while taking on the thankless task of dealing with the planning authorities in our attempts to broaden the farm and make it a more viable long-term proposition. Her decision to leave her TV work behind and come on board with the farm was huge. It meant the ideal combination of personalities – me forever wanting to jump in feet first, her imaginative and ambitious but focused and sensible. Michaela's also an incredibly strong person. While I was catching woodlice in Clavering, she was living barefoot in the Himalayas. Her late father had a military background and until she was seven his regiment was based in Nepal.

As I layer myself up before heading out, it would be tempting to think I don't like winter, that like many others I see it as cold and miserable, something to be endured before that welcome warmth of spring. The truth couldn't be more different. I treasure winter – the natural ending of the year. We've had the renewal of spring, the long, hot (hopefully) summer and autumn's falling of the leaves. Now it's the turn of the season of reset – a time of frosty mornings and lovely blue skies, open fires, Sunday lunches and afternoon walks.

It's a season, like late autumn, that concentrates your time, flat-out busy but then four o'clock comes and it's dark so you're forced in. Next day comes and out you walk again into one of those lovely crisp winter's mornings. Or, if the weather isn't quite so kind, a low grey sky and sleet! These days, the worst that winter can bring – as a kid in Essex, I remember climbing over massive snowdrifts – tends to miss us. There have been days where big snowfalls have cut us off and left us isolated, but they have been few and far between. A quirk of geography means that because we have the Orwell estuary nearby, a big heavy storm cloud will come straight at us only to swerve away right at the last minute. That means we've avoided major storm damage and the crazy high winds, like in the Great Storm of 1987, which can raze great swathes of woodland to the ground. Even so, when a storm does pass through, it's the woods I think of first. A lot of the time when a tree falls or is cut down, land-owners will chop them up and take them away. But if it's not in the way, I'm more minded just to let it rot. While modern woodland is managed, which is really beneficial for certain groups of animals and plants, traditionally, wild woods would be full of dead and decaying trees. When an oak goes down, just like when an elephant dies, there's a whole ecosystem that winds into action to break it down. A rotting log is a rich natural larder, as is dead standing wood, often perceived to be a hazard but which, as we have done with an old cherry tree, offers rich pickings if left safely to fall apart. There's a whole group of insects which feed on dead and decaying wood, and it's that very deterioration that means come spring we're rewarded with one of my favourite insects, stag beetles. They're rare or non-existent in parts of the UK, and yet our woodland is full of them, emerging after anything between three and seven years as larvae to spend their final few months searching for a mate. The stag beetle shows that every nook and cranny offers a potential habitat. Stop and look around and you will see nature's ingenuity with death – fungi growing on a damp old log, nesting birds collecting its wrapper of lichen and moss. In woodland, death doesn't mean an ending, it means regeneration.

The animals you really need to watch in a wildlife park are the ones that ordinarily live in a tropical climate where the temperature, balmy heat with high humidity, is fairly constant throughout the year. A tapir, for instance, is a natural Amazonian. The East Anglian climate is many things but never have I heard it described as that. Our damp, chilly autumns and freezing-cold winters really aren't a tapir's thing. It will put up with it but at the same time must be wondering what on earth is going on, a bit like if you booked a day at a spa and found it had only an ice-cold plunge pool.

To cheer the tapirs up, we'll give them a banana or two. Bananas are like magic wands – wave one in front of a tapir and instantly they change from super-friendly, chilled out, not fussed about anything, never in a hurry, to a frenzied bundle of 'Give it to me! I want it. Give me it!' Also, as with any animal more used to warmer climes, we'll make sure Tip Tap and Teddy have somewhere cosy and comfortable to retire to where they can heat up and get their slippers on.

Meerkats, native to sub-Saharan Africa, get the same treatment. We installed a heater in their enclosure that comes on if they cut a light beam. They soon learned how it operates and fashioned themselves a lovely little bit of Africa in the middle of winter, sitting back, exposing their tummies and sunbathing in front of it. Unfortunately, in the wild, heat can't be turned on and off. Climate change means the meerkats' natural habitat is heating up faster than they can adapt. Like many other animals that have spent millions of years adjusting to a specific environment, to survive they are now being required to do so in a matter of decades.

Even in the heart of winter, we still have new arrivals. It was mid-January when two armadillos, Polly and Preston, came to us from Longleat Safari Park. The six-banded variety, Polly and Preston can't curl up into a ball – that's the three-banded armadillo – but are no less cute for that, like little mini tanks, an armoured bundle of scales, bristles and claws, scuttling round, endlessly digging for food or digging burrows. Armadillos' eyesight isn't the best, so they rely on smell and touch to find their way through life.

Preston was a little overweight on arrival – it appeared his scales were struggling to contain him, a bit like a big bloke in the pub putting a bit of strain on his shirt buttons – and so straight away he found himself on a diet. The hope is that Polly will mate with him so he needs to tidy himself up if their home, the former meerkat enclosure, is to become a love-nest.

We opened the old meerkat bedrooms up for them, so they should at least be snug.

At the same time, we took delivery of two zebu, again a visually quite remarkable animal, essentially a mini-cow with a camel-like hump on the back and a large dewlap, a flap of skin which hangs below the neck. I was delighted to reacquaint myself with zebu, fond as I was of the pair they once had at Mole Hall. Hailing from Africa and the Indian sub-continent, the storage hump means they are extraordinarily well adapted to arid areas. Offering meat and milk, they have become a particularly popular cattle animal in such regions, able to contend with thermometer-popping temperatures – which they certainly didn't find in East Anglia in January!

For all the help we give the animals to fend off the cold, it's amazing the protection and resilience that nature provides. We might keep our turkeys in a lovely indoor space with food, water and straw laid on, but truth is that in the wild, turkeys live outside. Same with chickens – there are some incredible chicken houses out there, but actually in the wild, they'd be perched up in trees at night in all weathers. That's why they have feathers! Forget what we do here – nature has been dishing out the winter gear for millennia after millennia.

Many of our animals are kept in tropical conditions all year round, crocodiles included. So far removed from us as mammals and primates, the crocs are absolutely fascinating to me. Think about it, we can communicate with many mammals – dogs, for example – even if it's on a very basic level. With a crocodile, that connection is non-existent. They are intelligent and instinctive in a way that's totally removed from us, capable of switching from virtual torpor to high-intensity attack in the blink of an eye. Crocodiles are possibly the best example of how, no matter how much you think you know an animal, you should never underestimate it.

I once went to Crocosaurus Cove, in Darwin, which hosts some of the largest saltwater crocodiles in Australia. As part of the experience, slightly unnervingly, I was lowered into a croc-odile tank in a big Perspex tube. Food was then offered to tempt them over, the result being the next thing I knew an 18-foot saltwater crocodile was right up alongside me – as unsettling as it was remarkable. Only then did I understand the species' sheer majesty. Their size is remarkable.

Out of the water, they just seemed content to bask on a rock, like big fat slugs (I wasn't saying that to their faces, of course!). One keeper told me they weren't massively active any more, just waiting to be fed every day. 'But,' he pointed out, 'if you fell into that enclosure, the moment you hit the floor they would spring into life. A crocodile will sit there for twelve hours motionless, like it's dead, and then – bang!'

I knew immediately what he meant. I've always felt as a human that we can never really read crocodiles. I've met people who genuinely believe they can build that relationship with wild crocs, but in captivity their unpredictability should always be at the back of your mind. Think you know any animal inside out and all you're doing is lulling yourself into a false sense of security. It's not like if a crocodile gets hold of your arm you can smile and say, 'Come on now! Let go! Be a good boy!'

Naturally, we make sure any ranger who works with our crocs is trained to do so. Those who don't feel confident are never asked to deal with them. Ours are Morelet's crocodiles, a freshwater species, also known as Belize crocodiles, which average 3 metres in length. To manage them, we train them to respond to stimulus and reward. We have a long stick with a red ball, the idea being that they come to associate its presence with being fed. That way, when we need to persuade them to go into their holding closure to be weighed, we can bring out the ball and stick and they'll follow it. We use the same technique with the camels.

The tropical house is also home to our red-footed tortoises, native to South America and most familiar because of their shells, neatly patterned with light spots. In the wild, a red-foot digs burrows and tunnels up to 10 feet deep to escape extreme heat. Fortunately, they have no need here, otherwise we'd have to build new foundations. We also have African spurred tortoises, also known as sulcata tortoises, which live along the southern edge of the Sahara Desert. These are giants compared to the red-foots. Up to 2.5 feet long and weighing up to 50kg, they're beaten for size only by the giant tortoises, such as those of the Galapagos.

•

All our reptiles are rescue animals, including the tegu, a large South American black-and-white lizard with an incredible forked tongue. His prehistoric nature fascinates me, looking like something from a jerky old black-and-white adventure film as he moves stealthily around, flicking in and out that incredible forked tongue, which he uses to smell by collecting tiny particles in the air and then transferring them to his mouth. A big, powerful lizard, a tegu can grow up to 5 feet – you certainly wouldn't want to be nipped by one – and makes for a startling exhibit. But in an ideal world it simply wouldn't be here, it would be in a rainforest or roaming tropical grasslands. The reason the tegu is here is that it was saved from a pitiful fate after being kept domestically and used by those who train fighting dogs as bait.

Boa constrictors, like other large snakes, tend to end up as rescues simply because people run out of space to keep them. They buy them when they're small and pretty and look totally manageable, but of course the reality is they just keep growing and growing. A reticulated python, the world's longest snake, can grow to more than 16 feet. A fully developed anaconda, meanwhile, can have a girth of more than a foot. The largest, the green anaconda, can weigh in excess of 550lb. Measuring boa constrictors is one of the more unusual jobs I have undertaken here. Word of warning: start keeping large snakes and they'll soon see your cat or small dog as fair game. Snakes are mobile. They don't just lie in coils on the floor. They're arboreal, they want to climb up and around. It's often said the diagonal length of an enclosure should be at least the length of its occupant. Even with an 8-foot snake, that's a big space.

My girls aren't too keen on the boas. More than anything, I think it's about engagement. If a puppy looks at you with its head cocked to one side, there's a familiarity. You can have a pretty good stab at what it's thinking. Look a snake in the eye and, like with a croc, you haven't a clue. The girls aren't anti-reptile, though. Far from it. They've kept turtles, and one of them had a skink (a species of lizard with small legs), and then geckos after becoming obsessed with lizards while on holiday in Spain. They're a little like me. While many of the reptiles we have are non-native, because of their childhood interest, the girls are totally familiar with them.

Tropical environments aren't just about the big stuff, as we've tried to show with our new invertebrate enclosure, the incredible work of our ranger Tamzin. It's home to African train millipedes, the largest of their kind on Earth, often growing in excess of 30cm, as well as sun beetles, beautifully patterned in gold and brown and usually found amongst the leaf litter of the forest floors of the same continent. Conscious as ever of how the exhibit would be viewed by children, we installed a magnifying glass above the beetles' feeding area so they can watch as fruit, leaves and the culinary delight that is rotting wood are relentlessly devoured.

We offer the chance for people to adopt the animals. It doesn't mean they can take them home, it's more to do with helping to pay for their upkeep and supporting global conservation initiatives. I often feel sorry for our occupants who fail to be adopted. It reminds me of being picked for sports teams in the playground – it's always the same kids who are left until last. So while the meerkats and the anteater are never short of willing adopters, the likes of your millipedes, bearded dragons and stick insects are rather less likely to have a little plaque outside their enclosure revealing the name of their benefactor. Please, if ever you go to a wildlife park and the opportunity is there to sponsor a bullfrog, do it.

For a while, we had an aviary in the tropical house but were baffled when the birds started disappearing. We couldn't for the life of us work out what was happening, eventually installing CCTV cameras and the next day watching the footage back. Every now and then we saw flashes going up and down the tree branches, coinciding with the vanishing form of a zebra finch or fruit dove. You didn't need to be a genius to understand what was going on. Those flashes were a family of stoats. Clearly they had found a way into the building and were killing the birds. We tried to trap them, to cut off any potential points of entry, but in the end took the view that it was a futile battle. It's incredibly difficult to make a building stoat- and weasel-proof and sometimes, sadly, you have to admit defeat.

Those stoats are only doing what comes naturally to them after all, same as the rats we occasionally notice at this time of year when they start looking for indoor shelter and moving closer to potential food supplies. As somewhere with a lot of indoor shelter and plentiful food supplies, I can see how their minds work! They're not our only visitors. By now, young foxes, resplendent in rich new coats, have left their family to go it alone. Naturally inquisitive, there's a certain inevitability in them checking us out. Along with the locals, we have an influx of urban foxes from Ipswich, fat on takeaways snaffled from bins. Here they have to be a little more opportunistic if they're to bag one of our turkeys, but they still present a danger. As well as fencing, we do have another way to put them off – Jerry the alpaca. Our South American pal will quite happily live with turkeys and chase foxes off. We've even lent him out to guard turkeys at a farm nearby, receiving two rare-breed turkeys by way of return.

Tempting as it is, we can't put guard alpacas everywhere. With the chickens, we just have to lock them up and hope for the best. When a fox gets in with hens it can cause havoc, but it's a rare occurrence. I tend to think there's so many weird animal smells coming from our place – camels and tapir to name but two – that the average fox gives us a wide berth. It must be a little disconcerting to go from Suffolk to Mongolia and the Amazonian rainforest in a matter of footsteps, a journey marked out, in fox terms, predominantly by poo – a wildlife park means a lot of poo! Wheelbarrow after wheelbarrow is filled every day. And I love it. I'm not sure everyone at the park shares my enthusiasm, but as I zigzag through the various enclosures with my shovel, I'm consumed not by the need for a nose-clip but by how efficient these animals are, each and every one an incredible piece of evolutionary engineering, taking so much from their food with so little left behind. What they do leave behind is like gold dust. OK, you might not want to see a wodge of zebra dung in the display cabinet at your local jeweller's, but for a farm it's manna from heaven, the starting point of the perfect compost. Pile it up, a veritable feast for bacteria and bugs, give it a good fork every now and again to let the air in to break it down, and bingo, in a few months you're filling that same wheelbarrow with 24-carat compost. Not only that, but dung is a great clue to animal well-being, easily screened to check for parasite loads and other health indicators.

I could talk about dung a lot, but I understand it's not everybody's cup of tea and so perhaps it's best to return to the point in hand. While we don't want predatory natives in the park, I do love knowing that out there in the darkness the wildlife is doing its best to counter the threat of these hard months. Stoats, weasels, owls, badgers and the rest are all after the same bounty of small mammals such as wood mice and voles. The hardiness of animals in winter never fails to amaze me. I still find it incredible, when I step outside on a freezing-cold winter's morning, to think that the tiniest of animals survive below-zero temperatures night after night. Of course, unlike our wildlife park animals, none of which hibernate, plenty of the natives get their heads down for the winter, not just the hedgehogs and bats but reptiles and amphibians such as grass snakes and the common toad.

One winter, I fully plan to hibernate. 'Right,' I'll announce, 'I'm off to bed for three months.'

'Hang on!' the family will splutter. 'What do you mean?'

'Haven't you noticed?' I'll say. 'I've been eating five bowls of pasta, a Chinese takeaway and a roast dinner every day for the past month. I've put an extra duvet on the bed,' I'll tell them. 'See you in March.'

With that option disappointingly off the agenda, I live my hibernation vicariously. I particularly like walking into an old shed, looking up and seeing ten or fifteen peacock butterflies huddled together – a real 'Wow!' moment. You might even find a hibernating butterfly in your home, woken up early by the central heating kicking in. It's important not to put it out into the cold where it will fly around and use up all its energy. Much better to put it in a garage or shed so it can resume its hibernation.

It's not just central heating. Climate change means nature's winter thermostat is increasingly on too high. While we might not mind milder winters, they do lead to some animals coming out of hibernation when there's not enough food, their life cycles bent out of shape, often with disastrous consequences. It doesn't matter where you are, Suffolk or the Sahara, long-standing ecosystems are incredibly vulnerable to sudden change.

Other animals, such as the reindeer, are in their element in winter. Discarded in the summer to keep cool, their coats are now thick and luxurious. They look absolutely tip-top, just magical, almost as if the environment has changed to suit the animal rather than the other way round. I sometimes imagine the tapir looking on in envy. The reindeer and camels have these amazing winter wardrobes while Tip Tap and Teddy are standing there in pretty much nothing but their underpants. Reindeer are basically the equivalent of that person who has got all the best skiing gear, or the kid with the really cutting-edge sledge who all the other kids laugh at because there's been no snow for five years but, when the white stuff does come, is out there going twice as far twice as fast as everyone else.

Raccoons too grow a thicker coat, trapping body heat close to their skin. They love nothing more than to roll in the snow, while also tucking into as much food as they can get their paws on to tide them over until spring. Because they don't hibernate, raccoons are dependent on their fat reserves and in the wild can lose up to 50 per cent of their body weight in winter. In captivity, we can change their diet, dish out extra portions and up the calories – ice cream, chocolate gateau (actually, that might just be me). Whatever the animal, winter is really about delivering creature comforts, whether that's warm bedding or ensuring they have the correct type and amount of food to keep them going.

It can be baffling how one animal struggles with the cold when a close relation seems entirely unbothered. The coati is a member of the raccoon clan but isn't half as concerned with colder climes. Their natural habitat is a lovely warm forest floor in Mexico and yet they can sit in an English oak in November as if it's something they do all the time. Maybe they don't want anyone to think that they're suffering, like those football fans who stand naked from the waist up in January. It's worth remembering too that a lot of our animals would be unlikely ever to see snow in their natural habitat. The macaques are a case in point. They find it fascinating, like children, and also seem to understand that it's very fleeting. They want to make the most of it before it's gone.

Over on the farm, while the Suffolk Punches are extremely hardy, selectively bred to suit British weather patterns, we make sure they have lovely warm beds in the stable. The important thing with heavy horses is to keep them from totally trashing their paddock in the winter. The Suffolks have outside access but we try to restrict it so those big heavy feet don't chew up the land. One of my favourite mental escapes at this time of year is to head up to their field and watch as they come over, breath puffing out of their nostrils like steam trains. I might then head down to the woodland, so lovely and bright without all its foliage, making it easier to spot the squirrels and deer, while the evergreen species stand out like beacons.

I'll admit it, as darkness falls, I can get a little spooked in the woods. There's a story of a beast that lives amidst the trees. I know someone who used to play on the farm as a child. They were heading home one day when they saw a big hairy creature about 10 feet tall in the undergrowth. It scared them so much that they turned straight around and started running the other way – but there it was, still in the trees watching them. The kid was so scared that in the end he started going all the way around the woods to avoid any more spine-chilling encounters. I'm not really a believer in ghosts and ghouls. I'm not expecting our Saus-quatch to get up and start moving around anytime soon. Even so, I've been out with the dog sometimes and she's refused point-blank to go anywhere near the woods. When there's no one around and the branches are swaying, the light getting low, I'm quite glad she's given me an excuse not to go in there myself!

There is definitely something about woodland that lends itself to spookiness, that sixth-sense feeling of being watched. I've had that sensation a few times, only to look up and find a deer staring at me a few yards distant. At Mole Hall, I used to take the all-terrain vehicle and trailer up to the woods to collect branches to put into the marmoset and tamarin enclosures – they liked gnawing and scent-marking them. I was up there cutting branches on an early spring day, the bluebells just coming up, when all of a sudden the whole atmosphere changed. I felt a real sense of being overseen. There was an ancient lane nearby, once a well-travelled route, with an old gatekeeper's cottage and I felt sure whatever was happening was connected to that property. It was such a weird feeling – as if this whole woodland just didn't want me to be there. I wasn't going to argue. I was back in the vehicle and off in a flash.

On another occasion, the mystic Derek Acorah, sadly no longer with us, came to the farmhouse to investigate claims of paranormal activity. As part of a new show, we'd been filming in the property to examine how it could be renovated and once again put to good use. Strangely, the recording had shown up a number of orbs, small round floating spheres said by some to represent the spirits of the dead, although photography experts will tell you the phenomenon is better explained by airborne dust particles, of which there are clearly a lot in derelict old properties, being lit up by the camera. The director felt this could be evidence of some kind of other-worldly presence and thus Derek was called in to investigate, resulting in me and Dolly on the doorstep at midnight waiting for him to arrive and stage a séance. Dolly was adamant the whole thing was a load of old rubbish.

'All right,' I said. 'Do you want to go and wait in the house?'

'Oh, no,' he replied. 'Don't worry, I'll stay out here and keep you company.' Hmmm.

Derek turned up and didn't need long to be convinced we were in the presence of spirits, which was handy because it was getting cold and a tot of brandy would have gone down a treat. He identified the presence of an evil woman, Marjorie, locked in a room to die from consumption, and also a man whose son had perished in a terrible farm accident. The youngster, Derek told me, was forever with me, following me round the farm, standing behind me, watching as I went through my daily tasks. Initially, I thought that might be quite convenient – an extra pair of hands and all that – but then I remembered that's not how ghosts work. You don't often see a scythe or pot of raddle powder floating through the air on *Countryfile*.

By this time, arch-sceptic Dolly had become a massive believer, and it's true there was a sudden rush of cold air in the room which appeared to affect nothing but us – all the old cobwebs remained eerily still. In fact, Marjorie appeared to have become quite attached to Dolly, to the extent that Derek called upon Dolly to get his ghostly admirer to leave, directing her to the door. Whether she has since returned I'm unclear, but whatever the case, the farmhouse séance will always go down as one of the most bizarre nights of my life.

Back in the real world, over in our orchard, our fruit trees are looking a little lean after being pruned back. Hopefully we'll remember to label them, as without leaves it's hard to tell what they are. I love the fact that on a cold winter's day we can head inside and enjoy the jams and compotes that came from the fruits of the previous harvest – a taste of the summer in midwinter. This is the season for really relishing food. An old classic like a potato and leek pie really feels like fuel as well as something to tuck into and enjoy, as do casseroles, stews, curries, home-made loaves and, of course, Sunday roasts. You can even get away with claiming the same about a home-made rhubarb crumble slathered in custard.

I might also take the chance to dig over the vegetable beds, which, if I'm lucky, will prompt a visit from one of winter's most cheering elements – a robin. On a cold, damp day, put your fork in the ground and all of a sudden there it is perched on the handle – the most clichéd Christmas-card scene but one that is no less real for that fact. It seems totally out of kilter with nature for a tiny bird to trust the close presence of a human being, but we forget that actually this is a relation-ship not minutes but thousands of years old, dating back most probably to when hunter-gatherers were digging up roots, the robin following along in the knowledge that the disturbed soil would mean rich pickings. Making a garden friendship with a particular robin is as strange as it is heart-warmimg, but the truth is, the bird knows, just as its predecessors did, that here comes an easy meal, almost as if that intelligence has been passed down in its DNA. We have effectively been placed in its ecosystem, a connection between two species so apparently foreign that, against every rule of nature, actually works.

My problem with such moments of relaxation comes from the fact they are all too often punctured by an unfortunate love of symmetry. I like things to be ordered, preferably in rows, which I expect is one of the reasons entomology, the study of insects, and all the collecting and pinning it entails, appealed to me so much. A slab of British countryside goes against all that. My eye will always be drawn to the heavy branch hanging by a sliver or the bit of land that's suddenly decided to flood. You'd think I'd have learned by now – countering nature is an unwinnable battle. People are always wanting to put order into chaos, and nature in particular is always telling them what they can do with that order. Tangled it is and tangled it will remain. What I or anyone else wants really doesn't matter.

Same as the heavy horses, our cattle breeds are extremely hardy. The Galloways and Highlands can be out all year round no bother, as can the Dexters and British Whites. The reason we bring them in isn't necessarily about giving them a nice warm place to go. Again, as with the Suffolk Punches, it's about protecting the grazing, so, once the turkeys have said goodbye, we put them in the big polytunnel to ride out the rest of the winter. Before that can happen, though, there's a lot of mucking out to do – a hundred tonnes of turkey muck to be exact. After everywhere has been sanitised and cleaned, we construct a pen, put new straw down, and then in the cattle come. Clearly, turkeys and cows have little in common, but of all their many differences the most remarkable is in volume. Turkeys, especially those out enjoying the morning sun, make a hell of a racket. I'm not quite sure what they're saying, but they're obviously very keen on it being said. Such a contrast when, three weeks before Christmas, suddenly they're gone, replaced only by the occasional moo. It's like when the washing machine stops – only then do you realise how noisy it was.

For the rest of the year, the turkey tunnel is empty, at which point I turn into a mad professor and use it as my lab. I play around in there, experimenting with areas such as aquaponics, a food production system investigated by all kinds of cultures over the centuries, which combines keeping aquatic animals with growing plants, the latter thriving on the nitrate-rich food which the process delivers. I had a circular system going where I pumped water from a small fishpond up through the roots of vegetables, which filtered the fish poo for nitrates, and gave the fish clean water again.

I find that kind of experimentation hugely satisfying, while at the same time accepting that my childlike enthusiasm drives pretty much everyone else insane.

I'm lucky in that a lot of my TV work entails me visiting pioneers in farming and the keeping of animals, both in this country and around the world, hoovering up ideas and, no doubt to the irritation of all, returning home like a mad professor desperate to experiment here. I might go off to film in Mexico and come back with a load of weird seeds from a merchant I've bumped into over there. Or I'll be in Kenya and witness an incredibly different way of growing a crop, returning filled with vigour – 'I'm going to try that!' – only to be met by a load of groans and the odd muttered 'Oh God! Here we go again!'.

I always have little notepads on the go, drawing revolutionary new pig arks or scribbling down costings for a prototype sty. Problem is my daughters have a habit of getting hold of these pads. On one page will be a measurement for a piece of corrugated metal, on the next will be a drawing of Father Christmas. I'm sure the same thing will have happened to Leonardo da Vinci when he was drawing up plans for that helicopter.

Some of my tinkering is more reassuringly traditional. Often I'll take willow cuttings, pop them in water, and come spring have the pleasure of seeing them root, the beginnings of a whole new tree. It's true, nature really is remarkable, and sometimes you don't have to be slightly mad to prove it.

By now, Santa has arrived on the farm – only right and proper, what with his reindeer being here. Busy in his grotto, he doesn't have time to visit the animals, so we do that instead. The macaques especially enjoy unwrapping a box of toys, which might sound a little over the top but actually when any animal plays it is learning skills. Everyone receives a treat of some sort on top of their daily feed. Not your average stocking filler, the raccoons might tuck into a nice bit of minced beef. No one, so far, has requested a flatscreen television. In what can be a bleak old month, the festive season provides the warmest of glows, a lovely focal point for everyone to rally around, which is why sticking the Christmas jumper on and having a staff shindig is really important, a celebration not just of Christmas but the hard work and achievements of the year just gone.

Christmas Day and Boxing Day are the only two days we're shut to the public, but there are always staff on site to ensure each and every animal is looked after. I've worked on plenty of Christmas Days myself and, while I wouldn't ever want to miss that special time with my family, there have been times when I've looked across the farm and wildlife park, totally deserted in the icy December air, and felt so incredibly grateful for everything we've got. I'll be captivated by an egret, a beautiful ghostly white heron, which has flown over from the estuary and landed by the pond to see what fishy treats are available. Or find myself staring at the starlings in awe of their winter plumage, just phenomenal with its dashes of white among the iridescent green. If I'm really lucky, my old pal the robin will come and perch just feet away on a fencepost. I'm not sure what gifts will be awaiting me back at the house, but it's hard to think they could be better than this. Then I'll notice a gate hanging off its hinges and the moment is gone!

I'm a sucker for a good old-fashioned traditional Christmas. We'll have people over and I'll cook the dinner. My mum would be up at 5 a.m., putting pans of veg on to boil, every window steamed up for hours, but I try to be a bit more relaxed. After all, when you think about it, what's Christmas dinner but a big Sunday roast? I love the day itself because it's one of those times when no other rules apply. You can have a sherry at 10 a.m. and nobody bats an eye! I have a tradition of sticking a tie on for the walk up to the pub and then it's back for the big feed. The only sad thing about it is there's no longer the rush to get finished for the big adventure film in the afternoon. We'll have beef and ham over the festive period too, from our own animals, of course.

When it's all done and the decorations are put away for another year, it feels to me like winter is beginning its farewell, as if I'm packing it away alongside the baubles, the tinsel and the Santa's sleigh that we use on the farm. I feel excited, both at what the New Year will bring and that spring is on the way. Psychologically, it's a refresh; the slate has been wiped clean. Instead of cold mud and bone-chilling winds, there's pictures of lambs in my head. That was until we started recycling Christmas trees. Then all I could see was a huge brown mountain of dead wood.

•

It's the simple things that fill me with pleasure. It's why I like the routine of the seasons, same as I like the routine of home life. I like a school morning, the girls up and out on time, same as I like it when they come home and it's coats off, snack boxes in the kitchen, homework done and chilling out for the evening.

I think sometimes there's an expectation because you've appeared on the telly that you live some extravagant celebrity lifestyle, that you've been given some sort of magic key that unlocks a door into a world of bulging treasure chests where the only drink is Champagne. I've never seen that life any more than I've wanted to live it. And anyway, living among farmyard animals doesn't really fit the bill. Sows and sheep don't really qualify as showbiz.

It makes me laugh. I was once on the ITV morning show *Lorraine*, waiting in the green room with a couple of other guests and some high-up TV types when the pipe on a basin burst and water started pouring everywhere. Everyone went straight into panic mode, shouting for the maintenance team, declaring a major disaster, while I found a tea towel, tied it round the pipe and created a little barrier so the water couldn't escape. Honestly, the way people reacted you'd think I'd just created the cure for all known diseases. It's why I'd be no good on *I'm a Celebrity, Get Me Out of Here!* Everyone would be moaning about being starving hungry and I'd be down at the creek catching yabbies (an Australian crayfish) to eat.

I'm happy to be that person rooting about under a sink. For one thing, TV is a fickle beast and being 'normal' gives you a grounding for when it all goes away. For another, it stops you having some weird ego trip, thinking that somehow you're elevated above everyone else. It doesn't hurt every now and again to get a bit of dirt on your shoes.

I'm just not built in the traditional 'celebrity' way. It's the same with cars. My dad had vehicles ranging from a little red pick-up to an old Talbot, Renault 5 and a tiny Subaru van. He never hankered after some amazing head-turner of a car and I think that's been passed down to me. So long as I've got an old Land Rover, I'm happy. I can appreciate classic cars, but that's as far as it goes. I couldn't own one – all that kicking tyres and polishing wing mirrors isn't for me! I mean, a smart new car is lovely, but I always think, 'If you had that sort of money, why wouldn't you buy a great big greenhouse, fill it with tropical butterflies, and sit in it?' Your own private tropical paradise for the same price as an SUV. At the very least, I'd have a nice cow.

The truth of my life is it's not very glamorous. By the very nature of what I do it can't be. It's why you don't open *Hello!* magazine and see Hollywood actors with a wheelbarrow full of manure. Everyone's a product of how they grew up to some degree and I'm no different. My parents were very much hands-on, waste-not-want-not, make-do-and-mend. Even if it was subconsciously, by osmosis, chances are I was going to pick that up. In fact, I get a lot of pleasure from being like that, so much so that if I'm cooking Sunday lunch, I'll use the same water first to parboil my roast potatoes, then the carrots (I'll put them aside and heat them up again with some added butter and tarragon), and then the broccoli. Whatever's left I'll then use for the gravy. One pan of boiled water for the lot, saving gas and adding a whole lot of flavour in the process. And only one pot to wash up! Compare that to some of the chefs you see on TV. Yes, they make some amazing food but I always play a little game based around something the camera never reveals – how much washing-up?

I try as best I can to take that attitude into the park. Not only is reducing waste great for the environment but from a revenue point of view it's just as important as increasing your customer base. A saving is as important as a sale.

Eventually, as the year turns, snowdrops appear amidst the trees, the first sign of rebirth. Nowhere is nature's power of regeneration better illustrated than with woodland – those endless cycles happening right before your eyes, all triggered by the changing day length. With winter, it's tempting to think of a season of death and decay, but actually that leaf litter, now a mulch, was a gift from above for all the microorganisms on the woodland floor. Back in the autumn, those trees were essentially raining sustenance, and immediately those tiny creatures started working away to turn those leaves back into the plant food which will, when the seasons turn, once again power the great oaks to produce the acorns that so many animals, my pigs included, will feed on.

The end of any season is far from a finality – it's just the seeds of the beginning of something else.

CHICKEN STOCK

Makes 1.5 litres

After a Sunday roast, there's the question of what to do with the chicken carcass. The easy option would be to throw it in the bin. But to do so would not only have our grandparents' generation spinning in their graves but also deny us a world of flavoursome treats.

I always save the carcass. You're making the most of every bit of the bird and you're creating the basis of several more great meals to come – soup, gravy, stews, whatever you want. It's also a really good way of avoiding food waste and showing children a really useful skill.

Ingredients

1 onion
1 carrot
2 sticks of celery
1 leek
2 bay leaves
1 sprig of thyme
1 glass of white wine
1 chicken carcass
2 litres of water

Method

Start by chopping the veg into small chunks. Don't limit yourself – if you have parsnip or leek tops, for example, use them as well. Add all the veg, the herbs and wine, plus the chicken carcass and water, to a large pan. I even go so far as to encourage my children to scrape any leftovers off their plates – Yorkshire puddings, the lot – into the pan.

Bring to the boil, then reduce the heat, cover and simmer very gently for 1 hour. Strain into a large, heatproof bowl and allow the stock to cool.

Set aside all the veg, discard the herbs and pick any remaining meat off the carcass. At this point, beside me will be three bowls: one for the bones, one for the meat and one (with all the skin and the cartilaginous bits) for the dog!

Once the stock has cooled, any excess fat can be skimmed off and the remaining liquid strained through a fine sieve. I will then either add the remaining meat back to the stock or use it to make a delicious chunky soup by combining it with the reserved veg and a little stock.

Alternatively, use the stock immediately for something like a risotto. It can be stored in the fridge for four to five days, or frozen in an airtight, plastic container for up to 3 months.

The last few weeks of winter are all about tidying up ready for spring. The greenhouse has to be prepared for potting up, the butterfly house readied for the reintroduction of its incredible inhabitants, and everywhere checked over before the imminent influx of visitors. It's funny how our own burst of activity is mimicked by the reignition of life all around us. It's only early February and yet the toads, frogs and newts reappear, nature's calendar telling us quite clearly that an immense explosion of life is on the way – 'Get ready! Any minute now, it's about to start all over again.' Elsewhere, commercial lambing has started, indicated by the newborns in the fields, a sign that it surely won't be long before our own are on the way.

As much as I love winter, there's an undoubted internal boost from knowing that spring is tapping gently on the window-pane. I'll stand in the woodland and, as the sap begins to rise once again in the trees, it's as if a throbbing rhythm of energy can be heard louder and louder. If winter has done its job properly, it has wiped the slate clean, mercilessly blitzing the weak and benevolently sparing the strong. Everything, flora and fauna, is now ready once again to build its way up to full fruition. Included in that rebirth is us. Twenty times now we have boarded the runaway train that marks the start of spring and clung on by our fingertips as it rattles and weaves an unpredictable track back to its winter buffers. It's a seat-of-the-pants, occasionally white-knuckle journey, but those moments when the line straightens out and finally we can look up, breathe and take in our surroundings always make it worthwhile. Never once have we reached for the emergency stop button. Why would we? We know this annual trip through the seasons can never be bettered. For as long as we can haul ourselves aboard, every spring we'll be there, golden ticket in hand.

Even the Suffolk Punches are dwarfed by the JIMMY'S entrance sign – a leftover from a TV show, honest! With rangers Tom and Sophie.

Pigs did fly – Storm Eunice saw our giant piggy bank take to the air.

Reindeer have impressive headgear, and, unlike other deer species, both sexes sport antlers.

One of the great attributes of rare breeds is their hardiness.

Ice-spy – an emu inspects a winter landscape.

The peacock adds a bit of tropical warmth to the winter chill.

Feeding time with a friendly lemur, typically unruly meerkats, and Basil and his remarkable snout.

The Future

People sometimes ask me if, after twenty years, I still have the passion I had when I first set foot on the farm. There's an expectation that I might have tired of it after all this time, like you occasionally hear of a sportsperson who has fallen out of love with the game. But I can honestly state that the truth is exactly the opposite. There's so much variety that I could never get bored. In a single day, I might go from feeding the cattle to watching the pigs, growing giant pumpkins, watching a snake shed its skin, and bringing in a new troupe of macaques. Maybe if I'd kicked a football every day for twenty years I would be bored, but a farm and wildlife park? It's a dream (dotted with the occasional nightmare) that never ends.

Essentially, I'm a weathered version of the person who rocked up on that first day at Pannington Hall Farm and saw what could be. Back in 2002, I wanted to work hands-on with rare-breed animals. I wanted to do more than talk about conservation, I wanted to be part of it. I couldn't bear the thought that my life might pass me by without trying to make a difference – at the very least giving it a crack. Even if the entire enterprise had gone over a cliff after six months, I'd still have been able to look at myself in the mirror and know I'd had a go. I can do that now – the only problem being that the bloke who stares back at me isn't quite so youthful! In the same way, my energy levels have dipped somewhat and for some reason I tend to rip my jeans easier! That's the thing with getting older – you feel as if you're trapped in someone else's body. Only the other day, I went with Cora to buy her some tropical fish. The minute I entered the shop, I was back to being that little boy – 'Can I have one of those, please? Three of those? Two of these?' – my face up against the tanks. Then there was the excitement of them being bagged up, carrying them back to the car, putting them into the tank at home. Every emotion was still there. I'm the same now when the latest list comes in of animals that have been confiscated at the airport. I can't wait to have a look. There's undoubtedly a childlike element to that, a wonder at seeing what's available, inevitably tempered by the reality of what we can and can't realistically bring here.

I also still love the fact that, while the farm and wildlife park will always be hard work, it never loses its capacity to make me smile. Important, because now the venture is such a big operation, it's easy to get distracted and forget sometimes to get back to basics. I pull myself up sometimes because I walk around and see a rubbish bin that needs emptying instead of the incredible animal in the enclosure behind it. Then again, it's that truth that makes it interesting. I don't have a huge managed operation. I'm not some distant boss tucked away in his fancy office. I see every spider web in every dirty window and that's exactly how it should be. It's massively important to me that I never lose touch with what it is that made me want to do this in the first place. Being around animals is what I've always wanted. If you love the natural world, what could possibly be better than having your own wildlife park? Your new anteater heading down the M1 and his home not being finished is, admittedly, a little bit stressful, but dealing with the animals is what I love and why I do it.

When I started out with the dream of not just preserving some of the country's finest traditional farm animals but showing they can actually still fulfil a role in modern agriculture, I had no idea I'd become someone others looked to for advice. When my day-to-day life was one of mistakes and mishaps, endlessly chasing my backside, if anyone had told me that one day I'd be the longest-serving president of the Rare Breeds Survival Trust, a charity committed to saving our native farm animals, I'd have laughed in their face. If they'd added that none other than Prince Charles, a vocal supporter of the trust, would pop in to see how I was getting on, I'd have thought they'd eaten a dodgy mushroom.

But the truth is both those things have happened, same as those overgrown fields and rickety buildings are now home to a farm, wildlife park, shop and restaurant. You can even get married here! Michaela and I had our reception here, for goodness' sake. We took the journey back from the church not in a classic Jaguar or Roller but on the tractor I'd bought her for the garden – sprayed pink for the occasion, obviously. As if everyone in the area didn't already think I was out of my mind!

All that change has come about from trying to make the most of what we have, rather than wasting energy thinking about what we don't.

'What can we do here? Will it work?' Stop wasting time asking questions! There's only one way to find out! It's that attitude that has carried us through. To find out if you can swim, first you have to jump in the pool. That's how we ended up at one point with a theatre company in the woods. That tumbledown old building, which had seemed good only for the bulldozer, was suddenly alive with Shakespeare. It's how we ended up with 20,000 people at a festival, the Happy Mondays' Bez being perhaps the most unexpected exhibit ever at the park! It's how we ended up with an anteater and two tapirs.

More than anything, the park shows that escapism is something to cherish, not to dismiss. I'll never be immune to the fascination of any of our residents. I'm the master of the double-take. It's sad that our engagement with the natural world is so often packed away when we go to work, when life suddenly seems to get more serious – at the very point where connecting with nature would do us so much good. Children get it right when they immerse themselves in the natural world. If waiting excitedly for a consignment of ring-tailed lemurs to turn up makes me a bit of a kid, then I'm totally happy with that.

The child I was definitely still exists. Sometimes he gets pushed aside by the day-to-day of running a wildlife park, but he's always fighting to re-establish himself at the front of my mind. I see him now in my girls. They love the wildlife park and all want to be involved with it. It wouldn't be natural if Michaela and I didn't talk about legacy, what the future holds, how the park might carry on when we're no longer in a position to be quite so hands-on. I'm not a giant tortoise (although I'll probably end up looking like one), I can't go on for ever.

We would never push a life with animals on the children, but subconsciously there has to be a drip-drip effect. In fact, their first crate of animals arrived just this morning. When we were on holiday in Crete, they became very attached to a couple of rescue cats. Those cats have now joined us here – one of them greeting me with a bite. It's not just physical animals. Certainly, if we were still in a world of VHS, there'd be more than a few nature documentaries worn out by now. They're the same with books. I still have a lot of very detailed and specialised wildlife books going right back to when I was a kid at school. They'll grab one off the shelf and go through it looking for pictures of snakes from around the world, birds from the Amazon, all kinds of weird and wonderful stuff. I love seeing that. The internet is a wonderful learning tool, but there's something truly lovely about seeing a child open a book, turn a page, look at a picture and go 'Wow!' – like I must have done a thousand times, and might occasionally still do now!

Funny to think that soon enough it could be them doing the exploring. Maybe I'll be the one tagging along! I'm lucky to have travelled quite extensively, but there are still so many places I'd love to see. I find islands particularly alluring, the idea of habitats evolving in isolation, a fascination linked to reading about explorers, forgotten and hidden worlds, again when I was a kid.

I used to wonder how I'd get on if I was ever shipwrecked on the classic desert island. In the end, I concluded I'd be OK. I'd just collect all the different types of shells that washed up or record all the different plants I could find. The point is that all around us is endless interaction and fulfilment. Look at an old oak tree, how the ivy has grown around its branches, how the wildlife depends on its nooks and crannies, its intense harvest of acorns. If it's not nature's diversity that interests you, think about its architecture. Look into a foxglove, or a spider's web frosted silver on the ground. Take your pick of nature's offerings, there is always something fascinating. I could sit for hours in that environment. Perhaps it's that incredible variety on our doorstep that's meant I've never given in to the temptation to live abroad. Sri Lanka, with its combination of incredible landscapes and amazing wildlife, has always had a hold. We head out there every year as we've sponsored two orphans through the SOS Children's Villages charity and like to catch up with them and see how they're getting on. Every time we do so we're blown away by what a stunning country it is. I don't like leaving the farm for periods at a time, but travel can be good for delivering new ideas, a mental refresh, not to mention incredible perspective. I travelled back once from Kibera, one of the largest slums in Kenya, on the edge of Nairobi, after spending the day with a gang member who'd become an organic farmer. Barely was I off the plane than I was walking into the Glastonbury Festival. And there in front of me was a shanty town installation, corrugated iron formed into a haphazard mix of crumbling and shambolic structures, hundreds of well-to-do British kids dancing around eating lobster rolls and fully loaded fries, having paid the slum equivalent of a year's wage for a ticket. I couldn't help but wonder what the average Kibera resident would make of that. Were these privileged festival-goers making the connection?

Whenever we talk about a lifestyle 'choice', it's vital we recognise how lucky we are to live in our small percentage of the world. Veganism is a choice to us, but what kind of option is it to one of the billions who own no land; those who keep goats and cows because they have no other choice. They can hardly walk around with a bunch of carrots and replant them in every bit of soil where they rest up for a few days.

I once spent some time with the Maasai. There was a little lad, no older than seven, whose job it was to go out with the cows all day. He got up at six in the morning, had beans and rice, and then off he went, complete with a little club in his hand – in case of hyenas. Eight hours he was out there. I try to think of him when I get irritated because the wi-fi has gone down for ten minutes.

Writing this book has, in itself, given me perspective. Because the farm and wildlife park is constantly throwing up issues that need resolving, it can be quite a timeless place. Something you think happened two years ago turns out instead to have been seven. Michaela and I sometimes have to make ourselves take that step back, look around, and consider just what it is we've achieved together. More often than not, though, we're so involved with the day-to-day that it takes someone else to point it out. While each and every brushstroke adds up to a big picture, we're too close to see it like that.

What does make me stop and think occasionally is when I meet someone who hasn't been here for a decade or so. 'It's unrecognisable,' they tell me. 'It's changed so much.'

It's lovely to think just how many people have been on the journey with us. Equally, it's a little scary when they pull out old photos of the place, or remind us that this was once a dairy farm, or tell us their parents used to keep a certain rare breed. If they want to make Michaela and me feel really old, someone well into their thirties will tell us they used to watch *Jimmy's Farm* when they were a kid. Or that they were married here, which again makes us feel ancient when we see how old their children are!

I have to rein myself in a little bit sometimes, otherwise thinking about the people who have been part of this story, be it a line or a chapter, can feel a little overwhelming. When anyone comes here they add something to the narrative, and when they go they leave something as well. As has often been said, put your foot on wet sand and you always leave an impression. We all do it, more than perhaps we realise. We leave ideas, imprints, an intangible sense of something.

Going right back to when we started, I remember Roy who helped us out with the plumbing and electrics. Roy was a great character. Asa's granddad, he was of a generation where he constantly had a comb out, running it through his hair, the kettle always had to be on and payment was always in cash. He wore a boiler suit, under which he was never less than smart, and his van, even though it was probably forty years old, was immaculate. I love the fact that a lot of the work he did back then is still ensuring we're operational today. We were so thankful to Roy that we planted an oak tree complete with a little plaque with his name on it.

It's not just those who have worked with and helped us, there's people who have seen something come alive in their kids when they've been around the animals, those who have special memories, forged here, of loved ones sadly no longer with them. It's a place that's important for a million and one reasons, doubtless some that date from before we were even here.

Nowadays – and this is great to hear – there are a lot more people farming rare-breed pigs. I still get letters saying, 'We started our farm because of your show.' Or maybe they watched because it fitted in with an element of their agricultural studies.

Fantastic to see that the shows we make are reaching people and, as ever, entertaining rather than preaching along the way. We're not going to change the world in terms of producing rare-breed meat, but what we can do is show that rearing rare-breed livestock is viable and maybe offer inspiration for those thinking of going that way. It's more about being an amplifier of ideas, as opposed to trying to change the world one sausage at a time.

There is another thing that occasionally stops me in my tracks. In our early days, we planted more than a few saplings. They were finger thin. Now I can barely get both hands round their trunks, a lightning bolt reminder of how long we've been here and what we've done. Same when I stand in their shade.

'Hang on,' I'll think. 'Shade! When did that happen?' Again, with everything going on day-to-day, I hadn't twigged (if you'll forgive the expression!).

The orchard is one of the highest points on the farm. I look across the hills, pasture, bogs, fields and woods sometimes and actually can't help but laugh. 'How the hell did we actually do this?' It does feel like one minute we were sitting on a pile of rubble with a bottle of wine and the next we had three zebras out the back.

In all honesty, if I had my time again, I'd start with chickens and bees – slower, less pressured and more low-key than putting so much effort into being the archetypal British farm. Looking back, I fell into a slight trap of thinking, 'Pigs – tick.

292

Cows – tick. Sheep – tick.' Chickens would have fulfilled the desire to be free range while being less time-consuming and a lot easier to handle than livestock. I could also have saved time, food miles and money by slaughtering on site. Like James Bond, I'd have had a licence to kill. Quite literally – I'd have had it right there in my wallet, next to my slightly less glamorous Costco card (I expect 007 has one of those as well). It's not quite as simple as heading down to the post office and picking up a licence – there's a training process and the small matter of building a slaughterhouse – but birds are a much simpler commodity to handle than pigs, which require a journey to a fully blown red-meat slaughterhouse and the presence of a meat inspector. The cost is huge, to the extent that killing twenty pigs a week isn't really cost effective. Kill a few dozen chickens in your own little slaughterhouse and it's a whole different story, while customers appreciate the fact you are taking a more holistic approach.

With bees, people automatically think jars of honey. But there's a lot more to our buzzing friends than that. Yes, there's the honey we eat, but then there's also the honey that can be processed into mead. Honey changes with the seasons, which means different variations. Suddenly you can offer a variety of flavoured meads. Then there's propolis, a compound that bees produce from the sap of evergreens. Bees use this sticky substance in the construction of hives, for instance, to regulate ventilation at the entrance across the seasons. Incredibly, they will also use it to mummify an intruder. Mice are common raiders. If one should die in a hive, potentially after being attacked, the bees will entomb it in propolis to stop the spread of infection from the corpse. The antibacterial properties of propolis have been known for thousands of years. The Egyptians were big fans, taking a tip from the bees and using it in the mummy-embalming process. Nowadays its health benefits, not just antibacterial but antifungal and anti-inflammatory, are being taken on board by more and more health professionals.

The profit from bees wouldn't be huge – just like anything in farming! – but it's a way to add value to a commodity. Being able to produce your own brands is useful. Grow wheat and you are subject to the variations of the markets, just one tiny cog in a vast global mechanism. The whole thing is out of your control. With a product such as honey you can adapt. You are your own master.

Hindsight is a wonderful thing, but also, back in 2002, making sausages from rare-breed pigs was unusual. There were just a few of us doing it, which gave it a certain value. Over time, however, it became quite the norm, even popping up as a storyline on *The Archers*, the Radio 4 soap being a good litmus test of trends in farming. If it's happening down at Lower Loxley, then you know it's happening across the board.

That's not to say I'm riddled with regret. Absolutely not. I couldn't be happier with how these years have worked out. Sometimes, when everyone has gone home, I'll walk around on my own, revel in the isolation. I'll hear a peacock scream, see a tapir slide into the water, look up at the rooks swooping on the wind. It's almost a dreamlike state to be in, as if only the screeching of an alarm could possibly knock me out of it. Other times, in the early morning, I'll listen to the moorhens, lose myself in the dark shadows of fish in the pond. Life can feel quite complicated sometimes, but the farm is the anchor at the heart of it. I know I'm lucky to have that escape in nature, and, whenever I see kids in school groups taking everything in, seeing all these animals up close, maybe for the first time, learning about them and the importance they have in the world, I wonder if one day they might have the same experience.

I'm pleased to say there are actually a couple of schools near us which run their own farms as a micro-business. Every single pupil grows something, no matter what. One of the schools even has a chicken club, complete with chicken monitors who care for their flock, not just collecting eggs once a day but actually learning about animal husbandry – a great insight at such an early age – and how a farm runs as an enterprise, with eggs sold to a ready-made market of parents and supportive businesses. Everyone is desperate to be a chicken monitor – just as I would have been. It's such a simple idea, and has been a huge success, but actually, when you think about it, there would have been a time when pretty much every school would have had a few animals or an allotment. Funny how these old ideas are suddenly so relevant all over again. I'd love to see school farms become much more of a feature nationwide.

I'm lucky in that working on a wildlife park means my day is dotted with happiness, just as it was when I was at Mole Hall

all those years ago. It's strange the things that can give me pleasure. For instance, when we first started keeping cattle, which from the outside must seem quite innocuous compared to some of the more exotic animals, I remember being both overjoyed and really excited. It felt like something special, both to see cows back on the farm after such a long time and for me – who'd have thought it? – to be keeping them.

So much of my pleasure these days comes from the wildlife park's stability, like a sure-footed mountain goat, which allows us to plan for the future. To not only put a new adventure play area in for the kids but to do it with our own money was quite something. I remember once talking to a friend about business. He talked about the rough patches, the creditors knocking on the door, the periods of excruciating worry. That's what business life very often is. I know – we've experienced it ourselves. You can have a run of just breaking even for years, relying on every penny that comes in, effectively running on standstill. Should life then throw you a couple of curveballs, like the rainiest Easter on record, or the coldest summer, or maybe a vital bit of equipment packs up, then rapidly you find yourself in a financial nightmare. We don't want that unpredictability, we want to write a yearly budget based on solid figures, to know that if we invest a certain amount back into the business, we'll be able to buy a new tractor, build a new enclosure, or bring in that much-coveted Indian rhino. We'll be able to do the things that really make a difference, to the wildlife park, to the visitors, and to us.

I've always understood the value of things. I never received pocket money. I'd get a few quid for birthdays or Christmas, or a granddad would shove a pound coin in my palm, but otherwise money was hard earned. After washing plates covered in other people's leftovers from Sunday dinner at the pub, so pleased was I to end my shift that I'd ride my BMX home, knees hitting my chin, singing 'Born Free'! I knew also that if I smashed my aquarium, it would take me eight hours of such plate-washing to replace it. To me, a price tag indicated time as well as money. I appreciate that still.

Understanding the finances was without doubt our steepest learning curve. The old adage of 'turnover is vanity, profit is sanity' is absolutely correct. If there's no money in the bank, you can't feed the animals or pay the wages. You can't always be that student sticking your card in the cashpoint, resolutely ignoring your bank balance and squeezing another tenner out. OK, that tenner might appear but you're not looking at the ramifications. The real freedom comes from understanding the money, getting on top of it and being able to plan ahead. Not understanding it causes only chaos, worry and anxiety. Fail to learn and you soon become a busy fool. A fire-fighting busy fool – with an irate bank manager and bemused accountants. It's not an easy lesson but it needs learning, and quickly, if you are to build something sustainable. There are farms in this region that have 2,000 acres and are losing money. Because we're a tenth of their size, they look upon us as not a proper farm. Depends how you categorise it, really. Is a proper farm one that makes or loses money?

•

Sustainable we might be but never would you go into the rare-breeds and wildlife park business if you wanted to make a fortune. Feel free to have a poke in my garage – you won't find a Bentley. A business like this is built on a devotion to the animals and a love of the lifestyle. For anyone looking to go down this route, that twin ethos has to be there or you won't last five minutes. Just now I walked down and looked at the new lemur enclosure. As ever in a situation like that I felt excitement. That's what keeps me going – that next bit of the adventure. The last twenty years have been a big old slog at times, but that sense of adventure, that sense of 'How did this happen?', has never subsided. I can look at any part of the wildlife park and say with total honesty I'd never have dreamed of it being there when I first set foot on these acres.

Equally, I'm proud to say that never have we veered from the core values on which we built the business – to raise rare breeds in a way that is viable and, we hope, makes them satisfied and happy. The difference now is we're well established, but that doesn't make what we do any less involving or exciting. It's the ultimate truth of life that no one can predict what's around the corner. That's fine by me – I never came into this because I wanted a predictable life, in fact, entirely the opposite. There are many ways to introduce a seat-of-your-pants element to your life. Inviting crocs, anteaters, armadillos, macaques and meerkats to share their existence with you is one of the less orthodox, but I'd hazard a guess that of all the people to have tried it, very few (of those who have somehow managed to maintain their sanity) would say it's anything other than hugely satisfying and worthwhile – giving haircuts to guanacos aside.

Take a step – or in my case a leap – out of your comfort zone and it's amazing what can happen. A lot of people saw the Covid pandemic as a natural time to make that change, but really, what's to stop a person doing it at any other period of their life? If the time is right, the time is right. Look at Sir Tim Smit, who saw an old china clay pit in Cornwall and came up with the Eden Project. I've had the good fortune to meet Tim once or twice. I wanted to know how he'd succeeded with such an incredible plan, one that I'm sure many people would have looked at on paper and laughed at. The answer, it turned out, was nothing more than positivity, a belief that if you can convince the people conditioned to say no to say yes, just once, then anything is possible. Tim had a passion for plants, a belief in 'If you build it, they will come', and through sheer drive and positivity convinced others to follow. We're surrounded by animals that act on instinct. Sometimes it serves us well to act in that way too.

I know it's easy to say 'chase your dream', and I won't pretend that everyone who does so ends up in their own personal Valhalla – there was always a decent chance I'd end up back in a university lab pinning insects at some point – but there is something to be said about not living with regrets. The seasons themselves teach us a valuable lesson – that change is both natural and rewarding. It's the pattern of life, how our planet is set up. Sometimes it's worth turning your back on the bluster and the rush and listening to nature. Stare into the grass, see that mouse, that frog, that maybug, and you never know quite where they might take you.

Astonishing to think I've lived through eighty seasons – twenty springs, twenty summers, and so on – here now. Put it like that and it seems plenty, perhaps even as if an air of predictability might be sinking in. And yet nothing could be further from the truth. I might have travelled far and wide but I don't believe there's a country on Earth with a more beautiful array of seasons than this one right here. I don't believe there's anywhere on Earth where you so routinely pull back the curtains each morning not knowing whether you're going to be faced with the loveliest of rewards or the most intense of challenges. And for me personally, I don't see anything more important than working in the fields of animal diversity and conservation on an increasingly damaged planet.

I'm not saying there aren't times, when my hood's full of hailstones, there's a stone in my welly and I've been accidentally headbutted by a Suffolk Punch, that I don't yearn for a cup of tea and ten minutes with my feet up, but I can honestly say those moments are only ever fleeting. I went down this road because the thought of the nine-to-five, life passing me by, scared me. I saw that future staring me in the face. I perhaps didn't expect to replace it with a skunk staring me in the face, but I'll forever be glad I took the plunge.

•

It might be a few acres off the A14. It might be something people see fleetingly from the window of the 08.17 from Liverpool Street to Norwich. But to me, to us, Jimmy's Farm and Wildlife Park is a place that tells a wider story, that the natural world is our saviour. Look after it and we look after ourselves. That we can tell that story through an array of utterly incredible animals is, for me, the biggest of treats. Because remember, whatever you see me doing – sexing an emu, chasing a meerkat, filling a wheelbarrow in the camel enclosure – I am living my absolute dream.

Please – don't ever wake me up!

ACKNOWLEDGEMENTS

I am grateful to so many people for their help and support over the last twenty years.

Starting a farm or publishing a book isn't achievable on your own and both have been a huge team effort. I have been blessed to have had the help of some incredible people. Without their efforts and dedication, you wouldn't be reading this and I certainly wouldn't be at the farm today!

A massive thank you to my dear old friend Chris Terry for your fantastic photographs, endless jokes and reminding me when to pull in my double chin! To John Woodhouse, thank you for listening to my long stories and ramblings, you have a skill that can never be understated!

To Lindsey Evans and her fantastic team, Kathryn Allen, Louise Rothwell, Alara Delfosse and Lucy Hall at Headline – thank you for your expertise, patience and belief and for making this book a reality.

A huge thanks to Debbie, Danielle, Toni, Steph, Eloise, Ellis, Sophie, Tash and all the team at Fresh Partners.

To Stevie, Paul, Sven, Danny, Sophie, Tom, Sarah, Andrew and all my incredible team at Jimmy's Farm and Wildlife Park! Thank you for carrying the flag and fighting the good fight!

ACKNOWLEDGEMENTS

I've waited until the end to thank the most important person in my life and that is my wonderful wife, Michaela. I can't begin to thank you for your never-ending support over the last twenty years as we set off on our mad journey! Thank you for putting up with my ridiculous ideas and hair-brained schemes and sharing all the ups and downs that we have had. I wouldn't change a single moment.

Lastly to all the people who have taken time to visit our farm in Suffolk over the years – we couldn't have done it without you. You've quite simply made this adventure possible.

Here's to another twenty years!

INDEX